Collins

Student Support Materials for AQA

AS Physics

Unit 2: Mechanics, Materials and Waves

Author: Dave Kelly

William Collins's dream of knowledge for all began with the publication of his first book in 1819. A self-educated mill worker, he not only enriched millions of lives, but also founded a flourishing publishing house. Today, staying true to this spirit, Collins books are packed with inspiration, innovation and practical expertise. They place you at the centre of a world of possibility and give you exactly what you need to explore it.

Collins. Freedom to teach.

Published by Collins
An imprint of HarperCollinsPublishers
77-85 Fulham Palace Road
Hammersmith
London
W6 8JB

Browse the complete Collins catalogue at
www.collinseducation.com

© HarperCollinsPublishers Limited 2010

10 9 8 7 6 5 4 3

ISBN-13 978-0-00-734384-3

Dave Kelly asserts his moral right to be identified as the author of this work.

British Library Cataloguing in Publication Data. A Catalogue record for this publication is available from the British Library.

Thanks to John Avison and Stuart Jones for their contributions to the previous editions.

Commissioned and Project Managed by Letitia Luff
Edited and proofread by Jane Roth
Typesetting by Hedgehog Publishing
Cover design by Angela English
Index by Jane Henley
Production by Leonie Kellman
Printed and bound by L.E.G.O. S.p.A, Lavis (TN) - Italy

Contents

3.2.1 Mechanics

Scalars and vectors

Physical quantities can be classified into two groups: **scalars** or **vectors**. Scalar quantities, such as temperature or mass, have a magnitude (size) but have no direction associated with them. Scalars can be fully described by a single number and a unit. For example, stating that the room temperature is 20°C or that the mass of a person is 80 kg fully specifies these quantities. Other physical quantities, like velocity or force, have a direction associated with them. These are known as vector quantities. A vector quantity is only fully specified when the magnitude *and* the direction are given. It isn't sufficient to know that a force has a magnitude of 300 N; we also need to state what direction it acts in, e.g. a force of 300 N acting horizontally in a direction 30° east of north.

Examiners' Notes

If you are asked to find an unknown force or velocity, don't forget to give the direction as well as the magnitude.

Table 1
Examples of scalar and vector quantities met in this unit

Examiners' Notes

Make sure that you know which quantities are vectors and which are scalars.

Scalars	Vectors
distance	displacement
speed	velocity
energy	force
power	acceleration
mass	momentum

> **Definition**
>
> *A vector quantity has magnitude and direction, whereas a scalar quantity has magnitude only.*

Vector quantities are often identified by the use of **bold type**.

Adding vector quantities

When two vectors are added, we need to take account of their direction as well as their magnitude. Two vectors can be added by drawing a scale diagram showing the effect of one vector followed by the other, i.e. by drawing them 'nose to tail' (see Fig 1).

Fig 1
Adding vectors
The sum of **a** and **b** is found by drawing **a** and **b** so that the arrows showing their direction follow on

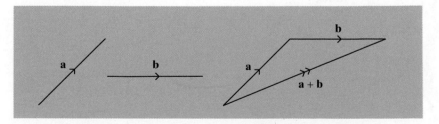

The sum of a number of vectors is known as the **resultant**. The resultant is the single vector that has the same effect as the combination of the other vectors. It is vital to take into account the relative direction of vectors when adding them together, for example the resultant of two 5 N forces could be anything from zero to 10 N, depending on their directions (see Fig 2).

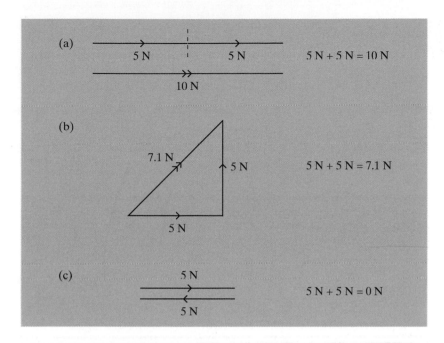

Fig 2
The magnitude and direction of the resultant depends on the orientation of the two component vectors

The resultant of two vectors can also be found by the **parallelogram law**. A parallelogram is constructed using the two vectors as adjacent sides. The resultant is the diagonal of the parallelogram (Fig 3).

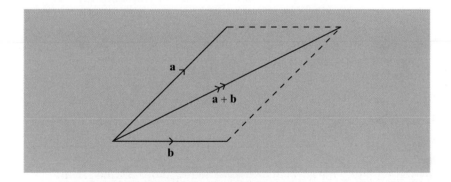

Fig 3
The parallelogram method for finding the resultant

If the vector diagram is drawn to scale, the resultant vector can be found by direct measurement from the diagram.

For two vectors at right angles, the magnitude of the resultant can also be found from calculation using Pythagoras' theorem (Fig 4).

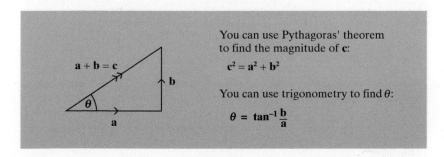

You can use Pythagoras' theorem to find the magnitude of **c**:

$$c^2 = a^2 + b^2$$

You can use trigonometry to find θ:

$$\theta = \tan^{-1}\frac{b}{a}$$

Examiners' Notes

The angle of the resultant for vectors at *any* angle to each other can be found using the sine rule and the cosine rule, but these methods are beyond the AS specification.

Fig 4
Adding vectors at right angles

Example

A tanker is being pulled into harbour by a tug boat which exerts a force of 200 MN in an easterly direction. The tanker is also subject to a force of 150 MN due to a northerly current. Find the resultant force acting on the tanker.

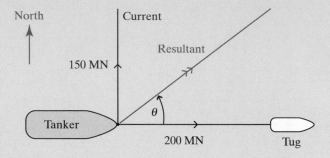

Answer

The magnitude of the resultant, R, is given by the equation:

$$R^2 = 200^2 + 150^2$$
$$= 40\ 000 + 22\ 500$$
$$= 62\ 500$$
$$R = 250 \text{ MN}$$

Find the direction of the resultant by:

$$\theta = \tan^{-1}\frac{150}{200}$$
$$= \tan^{-1} 0.75$$
$$= 37° \text{ north of east}$$

Fig 5
Subtracting a vector

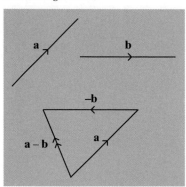

Subtracting a vector quantity can be thought of as adding a negative vector. The vector which is to be subtracted is reversed in direction. This reversed, or negative, vector is then added in the usual way (Fig 5).

Example

A river is flowing at $1\,\text{m s}^{-1}$. Find the speed and direction that a swimmer must travel if he is to achieve a resultant velocity of $1.5\,\text{m s}^{-1}$ directly across the river.

Answer

The swimmer's velocity, v, is the resultant minus the river's velocity.

Magnitude of velocity $= \sqrt{1.5^2 + 1^2}$
$= 1.8\,\text{m s}^{-1}$

Direction of velocity, $\theta = \tan^{-1}\frac{1}{1.5}$
$= 34°$

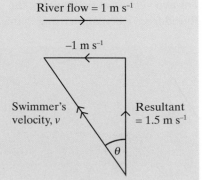

Resolution of vectors

A single vector can be replaced by a combination of two or more vectors that would have the same effect. This process is called **resolving** the vector into its **components** and it can be thought of as the reverse of finding the resultant. The components of a vector could be at any angle but it is often useful to use two components that are at right angles to each other. This might be to find the horizontal and vertical components of a force or a velocity (Fig 6).

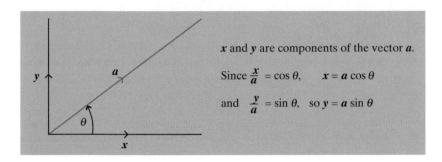

x and y are components of the vector a.

Since $\frac{x}{a} = \cos\theta,$ $x = a\cos\theta$

and $\frac{y}{a} = \sin\theta,$ so $y = a\sin\theta$

Fig 6
Calculating perpendicular components

Example

A wind is blowing at $15\,\mathrm{m\,s^{-1}}$ in a north-easterly direction. Find the components of the wind's velocity which blow towards the north and towards the east.

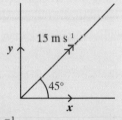

Answer

The northerly component, $y = 15\sin 45° = 10.6\,\mathrm{m\,s^{-1}}$

The easterly component, $x = 15\cos 45° = 10.6\,\mathrm{m\,s^{-1}}$

Example

A car of weight 10 000 N is parked on a steep hill which makes an angle of 20° to the horizontal. Resolve the car's weight into components that act along the slope and at 90° to the slope.

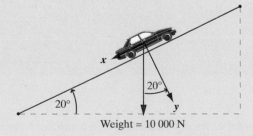

Weight = 10 000 N

Answer

Down the slope, component $x = 10\,000\sin 20° = 3420\,\mathrm{N}$

Perpendicular to the slope, component $y = 10\,000\cos 20°$

$= 9397\,\mathrm{N} = 9400\,\mathrm{N}$ (to 3 s.f.)

Essential Notes

Remember that the vector you are resolving is always the hypotenuse of a triangle. The components will always be smaller than the original vector.

Essential Notes

Gravitational field strength varies slightly from place to place on the Earth's surface, and decreases as you move away from the Earth.

Gravitational field strength on the Moon is about $1.6\,\text{N kg}^{-1}$, about one sixth of its value on Earth. On the Moon, you would weigh one sixth of your weight on Earth, though your mass would remain the same.

Gravitational field strength, measured in N kg^{-1}, is represented by the letter g. The symbol g is also used to represent the acceleration due to gravity (page 21) and can be given in m s^{-2}. These quantities are numerically equivalent.

Fig 7
For a regular shape, like a sphere or a cube, the centre of gravity is taken as the geometric centre. The centre of gravity of irregular objects depends on the arrangement and densities of materials in the object

Two or three coplanar forces acting at a point

It is often important to be able to identify, and add together, all the forces that are acting on an object. The size and direction of the resultant force will determine what happens to the object.

Everyday objects are subjected to a variety of forces, such as weight, contact forces, friction, tension, air resistance and buoyancy. All these forces, except weight, are electromagnetic in origin. They arise because of the attraction or repulsion of the charges in atoms.

Weight

This is the force that acts on a mass due to the gravitational attraction of the Earth. The gravitational field strength, g, on Earth is $9.81\,\text{N kg}^{-1}$. This means that each kilogram of mass is attracted towards the Earth with a force of 9.81 newtons. The weight of an object (in newtons) is given by:

weight (N) = mass (kg) × gravitational field strength (N kg^{-1}) $or\ W = mg$

The total weight of a real object is the sum of the gravitational attractions acting on every particle in the object. The resultant of all these forces is the weight of the object which can be treated as a single force acting at a single point in the object. This point is called the **centre of gravity**.

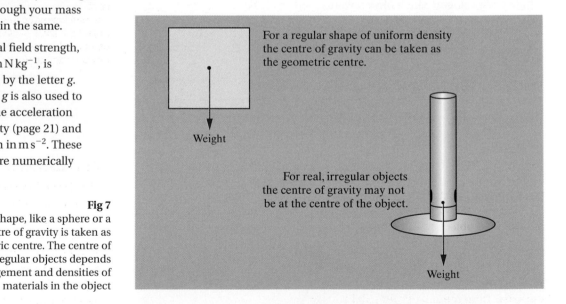

For a regular shape of uniform density the centre of gravity can be taken as the geometric centre.

Weight

For real, irregular objects the centre of gravity may not be at the centre of the object.

Weight

Contact forces

Whenever two solid surfaces touch, they exert a contact force on each other. This force is often known as the reaction. It is the contact force between the floor and your feet that stops gravity pulling you through the floor. The resultant contact force between two surfaces could be at any angle (Fig 8).

We usually split the contact force into two components:

- the normal contact force acting perpendicularly to the two surfaces
- the frictional force, acting parallel to the surfaces.

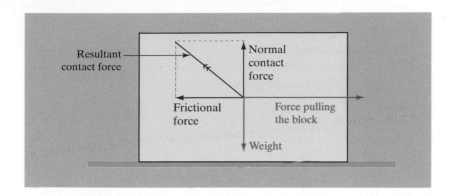

Fig 8
Forces on an object moving across a surface

Friction

A frictional force acts between two surfaces whenever there is relative motion between them, or when an external force is trying to slide them past each other.

Tension

An object is said to be in **tension** when a force is acting to stretch the object. Elastic materials, like ropes or metal cables, resist this stretching and exert a force on the bodies trying to stretch them.

Fig 9
The tension in the tow-rope, T, acts to pull the truck backwards and downwards, and to pull the car forwards and upwards

Air resistance

Any object that is moving through a fluid is subject to a resistive force or **drag**. Any object moving through the atmosphere has to push the air out of the way, this gives rise to the drag force that acts to oppose relative motion between the object and the fluid.

Buoyancy

Any objects that are partly or fully submerged in a fluid, like a boat floating on water or a hot-air balloon floating in the atmosphere, are subject to an **upthrust** from the surrounding fluid.

Free body diagrams

The forces acting on a real object may be quite complex. A **free body diagram** is an attempt to model the situation so that we can analyse the effect of the forces. The free body diagram is used to show all the external forces that are acting on an object. Since forces are vector quantities they are represented by arrows, drawn to scale and acting in the correct direction.

Essential Notes

The size of the air resistance acting on an object depends on the area of the object and on the density of the air. The air resistance also increases as the relative speed between the object and the air increases. So, as you go faster, the force trying to stop you increases.

Example

Draw a free body diagram for:

(i) A hot air balloon tethered by a cable.

(ii) A child sliding down a playground slide.

Answer

(i)

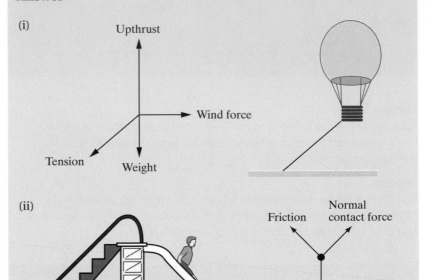

(ii)

Examiners' Notes

Make sure that you only include forces acting on the object you are considering, e.g. the child on the slide. Don't complicate things by including forces that act on other objects, e.g. the slide, or by including internal forces, like the tension in the child's muscles.

Fig 10
a + b + c must form a closed triangle if the body is to be in equilibrium

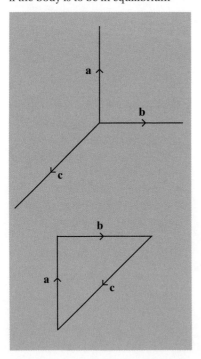

Equilibrium

Once we have identified all the forces acting on an object and drawn a free body diagram, we can use vector addition to find the resultant force. If the resultant force is not zero then the object will accelerate in the direction of the resultant force. If an object is not accelerating, it is said to be in **equilibrium**.

Definition

An object is said to be in equilibrium if it is stationary, or moving at constant velocity.

This idea is expressed in Newton's First Law of motion (see page 23). One of the conditions for equilibrium is that all the external forces that act upon the object must add up to zero. This means that if three forces are acting, the vector addition must form a closed triangle (Fig 10).

Example

A tightrope walker of mass 50 kg is standing on the middle of a tightrope. The rope makes an angle of 15° with the horizontal. Draw a free body diagram for the tightrope and find the tension in the rope. (To simplify the example, assume that the mass of the rope is negligible and take $g = 10\,\text{N}\,\text{kg}^{-1}$.)

Answer

Since the contact force between the rope and the person must balance the person's weight, the contact force, F, acting on the rope is 500 N. If the rope is to stay in equilibrium, the vector diagram must form a closed triangle.

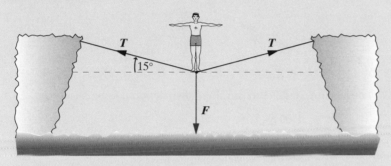

Free body diagram for the rope

$F = 500\,\text{N}$

Since the situation is symmetrical the tension, T, is given by:

$$\frac{250}{T} = \sin 15°$$

$$T = \frac{250}{\sin 15°} = 966\,\text{N}$$

An alternative method of investigating these problems is to resolve all the forces into two perpendicular directions, for example horizontal and vertical. If the object is to be in equilibrium, two conditions must be satisfied:

• the sum of the horizontal components must be zero
• the sum of the vertical components must be zero.

Example

A car of mass 1200 kg is parked on a hill inclined at 20° to the horizontal. The maximum frictional force between the tyres and the road is 5000 N. Will the car remain in equilibrium?

Answer

First we need to resolve the car's weight W into components that are perpendicular and parallel to the slope. For equilibrium perpendicular to the slope, the normal contact force R is:

$R = W \cos 20° = 12\,000 \times 0.94 = 11\,300\,\text{N}$

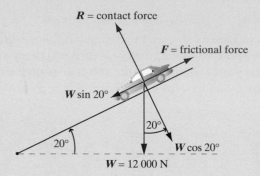

For equilibrium parallel to the slope the frictional force F is:

$F = W \sin 20° = 4100\,\text{N}$

This is less than the maximum value; the car will remain in equilibrium.

Moments

Moment of a force

Forces can cause objects to accelerate in a straight line. They can also have the effect of turning or tipping an object. Even when an object is acted upon by two equal and opposite forces, it may not be in equilibrium. If the forces do not pass through a single point, the object will tend to rotate. For example if you push a wardrobe at the top, and there is a large frictional force due to the carpet at the bottom, the wardrobe will tip rather than slide (Fig 11).

Fig 11
The moment about the point X causes the wardrobe to tip over

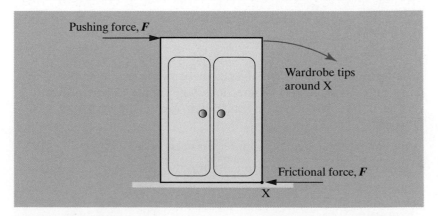

Fig 12
The moment of force F about X is given by
moment = Fs

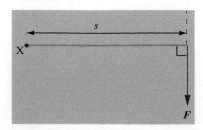

The turning effect of a force about a point is known as its **moment** or **torque**. The moment of a force about a point depends on two things:

- the magnitude of the force
- the perpendicular distance from the line of the force to the point.

Definition

The moment of a force about a point is equal to the magnitude of the force, F, multiplied by the perpendicular distance of the force from the pivot, s.

$$\text{moment (N m)} = F\ (\text{N}) \times s\ (\text{m})$$

The moment of a force is measured in newton metres.

You can increase the torque of a spanner on a nut by exerting a larger force, or by getting a longer spanner. When cycling a bicycle, the maximum turning effect is when the pedal crank is horizontal. When the pedal crank is vertical there is no moment, since the force passes through the pivot (Fig 13).

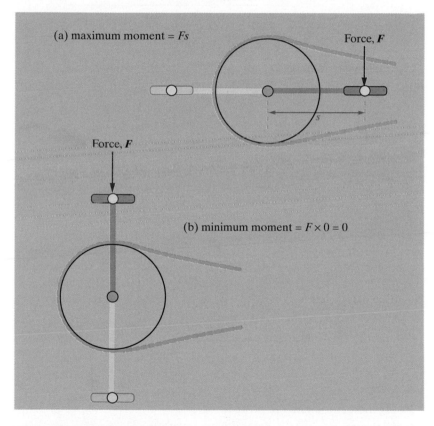

(a) maximum moment = *Fs*

Force, *F*

Force, *F*

(b) minimum moment = $F \times 0 = 0$

Fig 13
Moments on bicycle pedals

Essential Notes

It is important to note that it is the *perpendicular* distance from the pivot to the line of the force that is relevant. In Fig 14, where a vertical force is being used to lift a trap door, the moment is *Fs* cos θ.

Fig 14
Trap door hinged at X

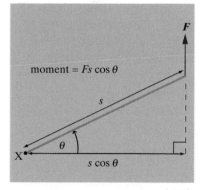

moment = $Fs \cos \theta$

The principle of moments

There is a second condition that must be satisfied if an object is to be in equilibrium under the action of several forces. Not only must the vector sum of the forces be zero, but the sum of the moments about any point (taking their directions into account) must be zero. If this condition is not met, the object would rotate around that point.

Definition

*The **principle of moments** states that if an object is in equilibrium the sum of the moments about any point must be zero.*

Another way of putting this is to say that the sum of the moments which tend to turn the object anticlockwise must be equal to the sum of the clockwise moments. The most straightforward example of this is on a see-saw. A heavier child can be balanced by a lighter child if the lighter child sits further from the pivot (Fig 15).

Fig 15
Equilibrium on a see-saw

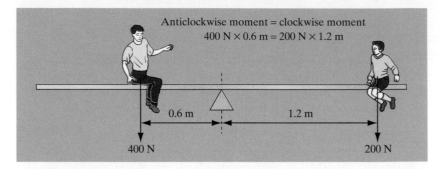

The principle of moments can be applied to find the magnitude of unknown forces.

Essential Notes

This lever gives you the ability to lift large weights with a smaller force. However, you will have to move your force much further than the slab will move. The **work** that you do will never be less than the **energy** gained by the slab.

Example

A crowbar (lever) is used to lift a paving slab which weighs 300 N. The crowbar pivots at a point 0.20 m from the slab. How much force will it take to lift the slab, if the force is applied 1.2 m away from the pivot?

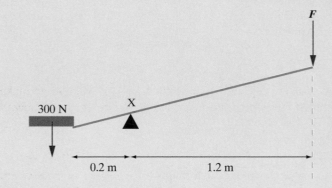

Answer

Taking moments about the pivot, X:

Anticlockwise moment = 300 N × 0.2 m = 60 N m

Clockwise moment = F × 1.2 m = 1.2 F N m

If the crowbar is in equilibrium these moments must balance:

$60 = 1.2\,F$, so $F = \dfrac{60}{1.2} = 50\,\text{N}$

To lift the slab, the force must be just greater than 50 N.

Sometimes it is necessary to apply *both* the conditions for equilibrium in order to calculate all the forces in a situation:

- The vector sum of the forces must be zero.
- The sum of the moments about any point must be zero.

Example

Two people are carrying a 3 m long plank which has a mass of 20 kg. Andrew is holding the plank at one end, whilst Beryl is holding the plank 1 m from the opposite end. Calculate the forces that each person must exert. (Take $g = 10 \, \text{N kg}^{-1}$ to simplify the example.)

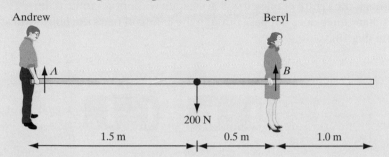

Andrew   Beryl

A B

200 N

1.5 m   0.5 m   1.0 m

Answer

For equilibrium, the sum of the vertical forces are equal: $A + B = 200 \, \text{N}$

Taking moments about Andrew's end of the plank:

Clockwise $200 \times 1.5 = 300 \, \text{N m}$

Anticlockwise $B \times 2 = 2B \, \text{N m}$

For equilibrium these must be equal: $2B = 300 \, \text{N m}$

So Beryl's force is 150 N. The rest of the 200 N plank is supported by Andrew, a force of 50 N.

Couples

Two parallel forces which act in opposite directions will tend to make an object rotate. If these forces are equal in magnitude, they are known as a **couple**.

Definition

The turning effect, or torque, of a couple is Fs, where F is the magnitude of one of the forces and s is the perpendicular distance between the forces (see Fig 16).

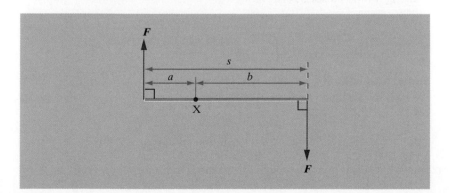

Fig 16
A couple. The torque of a couple about a point X is
$F \times a + F \times b = F(a+b)$
$= Fs$

Fig 17
A couple in an electric motor

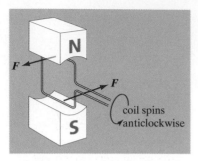

coil spins
anticlockwise

Essential Notes

In a place where gravity is uniform, the centre of mass and the centre of gravity (see page 8) are at the same point.

Fig 18
The applied force on the lower car does not pass through the centre of mass so the car spins

An example of a couple is in an electric motor where the force on each side of the coil is equal but opposite (Fig 17).

Centre of mass

All the mass of a body can be thought of as acting at a single point known as the **centre of mass** of a body. If the resultant force on an object passes through the centre of mass it will accelerate without rotating. If the resultant force does not pass through the centre of mass the object will spin (Fig 18).

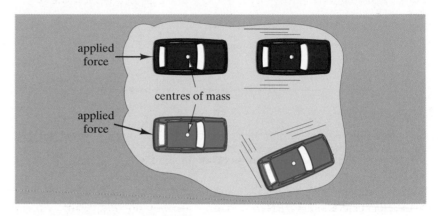

applied force

centres of mass

applied force

Motion along a straight line

Displacement and distance

Displacement and distance are both measured in the same units, metres, but displacement, *s*, is a vector quantity that describes the *effect* of a journey rather than the total distance travelled. Distance travelled is a scalar quantity.

Fig 19
The vector *s* represents the displacement. This is the net effect of the journey, and in this case has a much smaller magnitude than the distance travelled

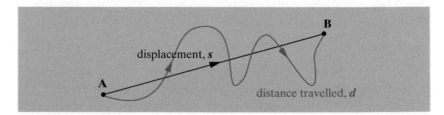

displacement, *s*

B

A

distance travelled, *d*

Definition

*Displacement, **s**, is the distance travelled in a given direction.*

Speed and velocity

Speed is the distance covered in unit time.

$$\text{speed} \, (\text{m s}^{-1}) = \frac{\text{distance travelled (m)}}{\text{time taken (s)}}$$

Speed is a scalar quantity which is measured in metres per second, or kilometres per hour. During a journey the speed may be changing. The average speed over the whole journey is given by:

$$\text{average speed} = \frac{\text{total distance covered}}{\text{total time taken}}$$

The speed at any given instant in the journey may be above or below the average speed. The speed at a certain time is known as the **instantaneous speed**. If we measure the distance covered in a very small time interval, Δt, the value for speed approaches the instantaneous value.

Velocity is a vector quantity; it has a magnitude (measured in $m\,s^{-1}$ or $km\,h^{-1}$) *and* a direction. Velocity is the speed in a given direction and is defined by:

$$\text{velocity (m s}^{-1}) = \frac{\text{displacement (m)}}{\text{time (s)}} \qquad v = \frac{\Delta s}{\Delta t}$$

Essential Notes

The greek letter Δ (delta) represents a change in a physical quantity; Δs is a small change in the displacement of an object.

Acceleration

Acceleration is the rate at which velocity changes.

$$\text{acceleration} = \frac{\text{change in velocity}}{\text{time taken for change}} = \frac{\Delta v}{\Delta t}$$

Since the change in velocity is measured in $m\,s^{-1}$, and time is measured in seconds, acceleration is measured in $m\,s^{-2}$. Acceleration is a vector quantity and therefore takes place in a particular direction. Any change in velocity, either speeding up, slowing down or simply changing direction, is an acceleration.

Acceleration does not always take place in the same direction as the velocity. A ball thrown in the air which rises and then falls again is always accelerating downwards due to gravity (Fig 20).

Essential Notes

A car driving round a roundabout may be travelling at a steady speed but it is constantly changing its velocity, because it is changing its direction.

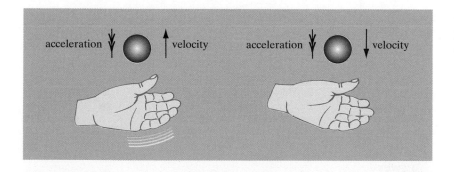

Fig 20
Acceleration and velocity of a ball thrown upwards (left) and falling back (right)

Example

A car goes from rest to 60 mph in 4.7 seconds. Calculate its acceleration. (1 mile = 1.6 km)

Answer

The car's final velocity is $\dfrac{(60 \times 1600)}{(60 \times 60)} = 27\,m\,s^{-1}$

$\text{acceleration} = \dfrac{\Delta v}{\Delta t} = \dfrac{27}{4.7} = 5.7\,m\,s^{-2}$

Displacement–time graphs

A journey can be represented by a graph showing displacement against time. The gradient of the graph represents the displacement in a certain time interval, which is the velocity. A straight line represents constant velocity.

Definition

*The gradient of a displacement–time graph is the **instantaneous velocity**.*

Fig 21
Instantaneous velocity and average velocity

(i) The instantaneous velocity at time t is the gradient of the curve at that point, $\Delta s/\Delta t$

(ii) The average velocity for the whole journey is the gradient of the straight line drawn from A to B, S/T

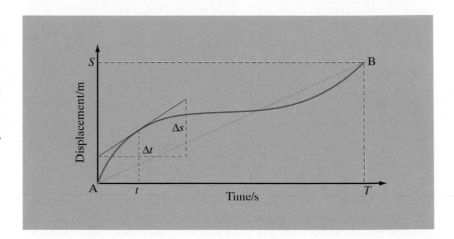

Example

The displacement–time graph below shows the motion of a car over 10 seconds. Describe what is happening at each stage of the journey.

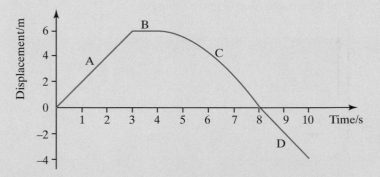

Answer

A Car moves with a constant velocity of $\dfrac{6}{3} = 2\,\mathrm{m\,s^{-1}}$

B Between 3 and 4 seconds, the car is stationary.

C The car returns to its original position with non-uniform velocity. The car moves slowly at first and then more quickly.

D The car moves in the opposite direction (negative displacement) with a uniform velocity of
$\dfrac{4}{2} = 2\,\mathrm{m\,s^{-1}}$

Velocity–time graphs

A velocity–time graph for a journey can be used to calculate the acceleration and the displacement. The gradient of the graph is $\Delta v/\Delta t$, which is the instantaneous acceleration. A straight line represents constant acceleration. A line with a negative gradient represents a negative acceleration. This could be slowing down (deceleration or retardation) *or* it could mean that the object is speeding up in the opposite direction.

The area below the velocity–time graph gives the total displacement.

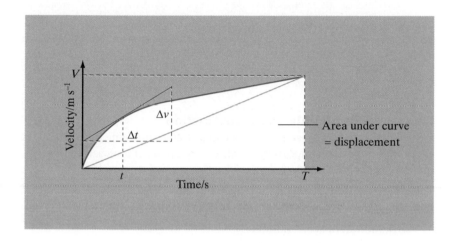

Fig 22
Velocity–time graph showing how to find the acceleration and displacement. The instantaneous acceleration at time t is given by the gradient $\Delta v/\Delta t$.

The average acceleration over the time T is the gradient of the straight line, V/T.

The displacement is the total area under the curve

Example

The velocity–time graph below shows a person's journey on foot.
(i) Describe each section of the journey as fully as possible.
(ii) Calculate the displacement during the first 7 seconds.

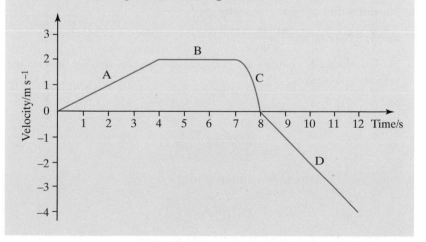

Answer

(i) A Uniform acceleration of $\dfrac{2}{4} = 0.5\,\mathrm{m\,s^{-2}}$

    B Constant velocity of $2\,\mathrm{m\,s^{-1}}$

    C Slowing down (negative acceleration) but at an increasing rate.

    D Negative uniform acceleration: the person is now speeding up in the opposite direction.

       Acceleration $= -\dfrac{4}{4} = -1\,\mathrm{m\,s^{-2}}$

(ii) Displacement is the area under the graph $= \left(\dfrac{1}{2} \times 4 \times 2\right) + (3 \times 2)$

$$= 10\,\mathrm{m}$$

Table 2
Variables in uniform motion

Quantity	Symbol
displacement	s
initial velocity	u
final velocity	v
acceleration	a
time	t

Equations of uniformly accelerated motion

We will just consider objects that move in a straight line with uniform acceleration. The five important variables that are used to describe this motion are shown in Table 2.

There are a number of equations which link these variables together and describe uniformly accelerated, straight-line motion.

1 The definition of acceleration is:

$$\text{acceleration} = \frac{\text{change in velocity}}{\text{time}} \quad \text{or} \quad a = \frac{(v - u)}{t}$$

Rearranging this gives:

$$v = u + at$$

2 The definition of average velocity is:

$$\text{average velocity} = \frac{\text{displacement}}{\text{time}}$$

But if the velocity changes at a constant rate we can say that the average velocity is $(v + u)/2$.

So

$$\frac{(v + u)}{2} \times t = s$$

or

$$s = \tfrac{1}{2}(u + v)t$$

3 Equations 1 and 2 can be combined to give:

$$s = ut + \tfrac{1}{2}at^2$$

4 Equations 3 and 1 can be combined to eliminate t:

$$v^2 = u^2 + 2as$$

These four equations can be used to solve problems about motion.

Examiners' Notes

Some people refer to these equations as the 'suvat' equations. It is a good idea to start each problem by writing down 'suvat'. Then you should identify which of the variables you have values for, and which equation you need to use.

Example

A sprinter accelerates from rest to $11\,\mathrm{m\,s^{-1}}$ in the first 4 seconds of a race. Assuming that his acceleration is constant, find the acceleration and the distance covered in the first 4 seconds.

Answer

$$s = ? \qquad u = 0\,\mathrm{m\,s^{-1}} \qquad v = 11\,\mathrm{m\,s^{-1}} \qquad a = ? \qquad t = 4\,\mathrm{s}$$

To find acceleration we can use the equation:

$$v = u + at; \quad a = \frac{v - u}{t} = \frac{11}{4} = 2.75\,\mathrm{m\,s^{-2}}$$

To find the displacement we can use $s = \frac{1}{2}(u + v)t$

$$s = \frac{1}{2} \times (0 + 11) \times 4 = 22\,\mathrm{m}$$

Acceleration due to gravity

A falling object accelerates towards the Earth because of the Earth's gravity. The **acceleration due to gravity** is independent of the object's mass. Experiments show that on Earth all objects, regardless of mass, accelerate under gravity at $9.81\,\mathrm{m\,s^{-2}}$.

A falling object, dropped from an aircraft, for example, soon reaches a high velocity. After 1 second it will be travelling at about $9.8\,\mathrm{m\,s^{-1}}$, and after 10 seconds it will have reached about $98\,\mathrm{m\,s^{-1}}$. If no other forces were acting the object would keep accelerating until it hit the Earth's surface. However, on Earth there is always some air resistance which opposes the motion of the falling object. Air resistance, or drag, has little effect on compact objects moving at low speeds, but the drag force increases with speed and eventually it is equal to the gravitational force pulling the object towards Earth. When this happens there will be no more acceleration and the object will continue to fall at a constant velocity, known as the **terminal velocity** (Fig 23).

Essential Notes

The acceleration due to gravity varies slightly from place to place on the Earth's surface; for example, it is greater at the poles than at the equator.

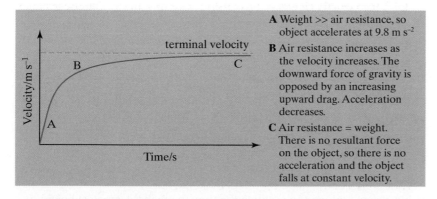

A Weight >> air resistance, so object accelerates at $9.8\,\mathrm{m\,s^{-2}}$

B Air resistance increases as the velocity increases. The downward force of gravity is opposed by an increasing upward drag. Acceleration decreases.

C Air resistance = weight. There is no resultant force on the object, so there is no acceleration and the object falls at constant velocity.

Fig 23
Velocity–time graph for an object falling in the Earth's atmosphere

Essential Notes

Cars have a top speed because the drag due to air resistance and other resistive forces increase with speed. When the drag force is equal to the maximum driving force that a car can produce, the car will no longer accelerate. It has reached its top speed.

Terminal velocity depends on the surface area of the object, as well as its mass, and on the density of the air. A falling feather floats down at less than $1\,\mathrm{m\,s^{-1}}$ whilst a free-fall parachutist may reach a terminal velocity of $65\,\mathrm{m\,s^{-1}}$ (close to 150 mph).

Fig 24
An object dropped vertically and one thrown horizontally will fall at the same rate

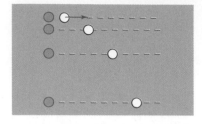

Projectile motion

An object thrown through the air follows a parabolic path. Even though this is not a straight line, we can still use the equations of motion. This is because horizontal motion does not affect vertical motion (see Fig 24).

This means that a two-dimensional problem can be solved by treating it as two one-dimensional problems, i.e. keeping the horizontal and vertical motions separate.

For example, if an object is thrown with initial speed $20\,\mathrm{m\,s^{-1}}$ at an angle of 45° to the horizontal, we can calculate its horizontal range and the maximum height it will reach.

First we resolve the initial velocity into horizontal and vertical components (Fig 25).

Fig 25
Components of velocity for a projectile

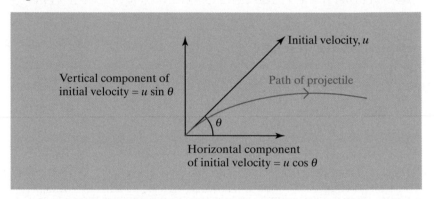

Vertical component of initial velocity $= u \sin \theta$

Initial velocity, u

Path of projectile

θ

Horizontal component of initial velocity $= u \cos \theta$

Examiners' Notes

Because displacement, velocity and acceleration are all vector quantities it is important to decide on a sign convention. When you start a problem you need to decide whether up is positive or negative and then stick to this throughout the problem.

If air resistance can be ignored, there is no horizontal acceleration, so the initial and final horizontal velocities are the same.

The variables s, u, v, a and t for the horizontal and the vertical motion of the object are as shown in table 3.

Table 3
Variables in the horizontal and vertical motions of a projectile

Horizontal motion	Vertical motion
s	s
$u = 20 \cos 45°\,\mathrm{m\,s^{-1}} = 14.1\,\mathrm{m\,s^{-1}}$	$u = 20 \sin 45° = 14.1\,\mathrm{m\,s^{-1}}$
$v = 20 \cos 45°\,\mathrm{m\,s^{-1}} = 14.1\,\mathrm{m\,s^{-1}}$	v
$a = 0\,\mathrm{m\,s^{-2}}$	$a = -9.81\,\mathrm{m\,s^{-2}}$ (down is negative)
t	t

Consider the vertical velocity first. When the object reaches its greatest height its vertical velocity will be zero: $v = 0\,\mathrm{m\,s^{-1}}$.

We can find the greatest height s using $v^2 = u^2 + 2as$:

$$s = \frac{(v^2 - u^2)}{2a} = \frac{-200}{-19.6} = 10.2\,\mathrm{m}$$

Using $v = u + at$, we can evaluate the time to reach the greatest height, t:

$$t = \frac{(v - u)}{a}$$

$$t = \frac{-14.1}{-9.81} = 1.44\,\mathrm{s}$$

Then consider the horizontal motion. The time of flight is twice the time taken to reach the greatest height, so $t = 2 \times 1.44 = 2.88\,\text{s}$. The horizontal range is given by:

$$s = ut = 14.1 \times 2.88 = 40.6\,\text{m}$$

Newton's laws of motion

Newton's laws of motion were published in his *Principia* in 1687. His work on mechanics built on that of Galileo.

Newton's First Law

> **Definition**
>
> *Newton's First Law of motion states that every object will continue to move with uniform velocity unless it is acted upon by a resultant external force.*

This law expresses the idea that objects will stay at rest, or keep moving in a straight line at a steady speed, unless an external force acts on them. The law restates Galileo's law of inertia. **Inertia** is the reluctance of an object at rest to start moving, and its tendency to keep moving once it has started. This law isn't immediately apparent on Earth. If you give an object a push, it doesn't keep going in a straight line for ever, because on Earth it is difficult to avoid external forces like gravity or friction. In space, well away from any gravitational attractions, objects just keep moving in a straight line.

Another way of stating Newton's First Law is to say that an object will remain in equilibrium, unless it is acted upon by an external force.

Newton's Second Law

Newton's Second Law relates to an object's **momentum**. The momentum of a moving object is defined as its mass multiplied by its velocity. Momentum is a vector quantity, so its direction is the same as its velocity. It is given the symbol p. There is no special name for the unit of momentum; its units are those of mass $\times$ velocity, $\text{kg}\,\text{m}\,\text{s}^{-1}$.

> **Definition**
>
> *Momentum $(\text{kg}\,\text{m}\,\text{s}^{-1})$ = mass (kg) $\times$ velocity $(\text{m}\,\text{s}^{-1})$ or $p = mv$*

The momentum of a body is a measure of how difficult it is to stop it. A heavy lorry could have a mass of 40 tonnes. When it is travelling on the motorway at $25\,\text{m}\,\text{s}^{-1}$, the momentum is:

$$p = 40 \times 10^3\,\text{kg} \times 25\,\text{m}\,\text{s}^{-1} = 1.0 \times 10^6\,\text{kg}\,\text{m}\,\text{s}^{-1}$$

Compare this to the momentum of a person running at top speed:

$$p = 80\,\text{kg} \times 10\,\text{m}\,\text{s}^{-1} = 800\,\text{kg}\,\text{m}\,\text{s}^{-1}$$

Fig 26
If the lift is to keep moving down at a steady velocity, by Newton's First Law its weight must be balanced by the friction, F, and the tension in the lift cable, T. For equilibrium, $W = F + T$

Essential Notes

Momentum can also be given in units of newton seconds $(\text{N}\,\text{s})$; this is exactly equivalent to $\text{kg}\,\text{m}\,\text{s}^{-1}$.

Using the concept of momentum, Newton's Second Law states the effect of a force on the motion of an object.

> **Definition**
>
> *Newton's Second Law of motion states that the rate of change of an object's linear momentum is directly proportional to the resultant external force. The change in momentum takes place in the direction of the force.*

This law defines a force as something that changes an object's momentum. Force, F, is proportional to the change in momentum, p, divided by the time taken for the change, Δt.

$$F \propto \frac{\Delta p}{\Delta t} \ \text{ or } \ F \propto \frac{\Delta(mv)}{\Delta t}$$

This can be written $F = \dfrac{k\,\Delta(mv)}{t}$, where k is a constant.

If the mass of an object does not change then m is constant and:

$$F = \frac{km\,\Delta v}{\Delta t}$$

but $\dfrac{\Delta v}{\Delta t}$ = acceleration, a, so

$$F = kma$$

The unit of force, the newton (N), is defined to be equal to 1 when $m = 1\,\text{kg}$ and $a = 1\,\text{m}\,\text{s}^{-2}$.

> **Definition**
>
> *The S.I. unit of force is the **newton**, N. One newton is the force that will accelerate a mass of $1\,kg$ by $1\,m\,s^{-2}$.*

Essential Notes

$F = ma$ holds only for a constant mass.

This means that in $F = kma$, $k = 1$ and we can write:

$$F = ma$$

force (N) = mass (kg) $\times$ acceleration $(\text{m}\,\text{s}^{-2})$

This way of writing Newton's Second Law allows us to calculate the effect of a force on an object.

Example

The total mass of a lift and its passengers is 1000 kg. The tension in the cable pulling the lift up is 12 000 N. Find the acceleration of the lift. (Ignore friction and take $g = 10\,\text{N}\,\text{kg}^{-1}$ to simplify the example.)

Answer

Upward force = 12 000 N

Downward force $- mg = 1000\,\text{kg} \times 10\,\text{N}\,\text{kg}^{-1} = 10\,000\,\text{N}$

Resultant upwards force = 2000 N

Using $F = ma$, $a = \dfrac{2000}{1000} = 2.0\,\text{m}\,\text{s}^{-2}$

Examiners' Notes

Remember to find the resultant force first, then use Newton's Second Law to find the acceleration.

Newton's Second Law also explains why it is advisable for a car to decelerate relatively slowly, since stopping in a short time can lead to high forces. Seat belts, air-bags and crumple zones in cars are all ways of increasing the time taken to come to a stop during a crash. This reduces the force.

In some cases, such as rockets and jets, mass cannot be treated as constant. It is then necessary to think of force as the rate of change of momentum, and to apply Newton's Second Law in the form:

$$F = \frac{\Delta(mv)}{\Delta t}$$

Examiners' Notes

You will not be asked a question involving changing mass at AS. This will be dealt with further in Unit 4.

Newton's Third Law

Definition

Newton's Third Law of motion states that if an object, A, exerts a force on a second object, B, then B exerts an equal but opposite force back on object A.

This law means that force between two bodies always acts equally on both objects, though in opposite directions. When we say that someone weighs 500 N, we mean that the gravitational attraction of the Earth on the person is 500 N. The person also attracts the Earth upwards with a force of 500 N. When a car exerts a force on the ground through the friction between its tyres and the road surface, the car pushes the ground backwards, whilst the ground pushes the car forwards. The two forces involved in Newton's Third Law never cancel each other out, because they act on different bodies.

Essential Notes

This law is sometimes written as 'every action has an equal but opposite reaction', but you need to be careful with this statement. Remember that *the forces act on different bodies.*

The two forces in Newton's Third Law are always of the same type, e.g. gravitational.

Example

A man stands still on the surface of the Earth. Draw the forces acting on the man and on the Earth and explain which forces are equal to each other and why.

Answer

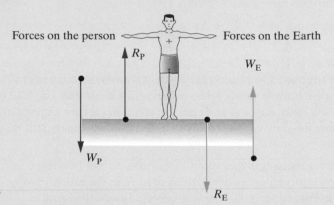

Forces on the person Forces on the Earth

Fig 27

$W_p = W_E$	By Newton's Third Law, the weight of the person is equal to the gravitational attraction of the person acting on the Earth.
$R_p = R_E$	By Newton's Third Law, the contact force of the Earth pushing on the person is equal to the contact force of the person pushing on the Earth.
$W_p = R_p$	By Newton's First Law, because the person is in equilibrium, the person's weight is balanced by the contact force of the Earth.
$W_E = R_E$	By Newton's First Law, because the Earth is in equilibrium, the gravitational attraction of the person is balanced by the contact force of the person on the Earth.

Fig 28

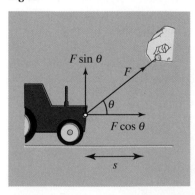

Work, energy and power

Work

Work is done whenever a force moves through a distance in the direction of the force. A force that does not move does no work. Sometimes the motion is not in the same direction as the applied force, e.g. when a toy tractor is being pulled along by a string (Fig 28).

The force, F, can be resolved into two components: parallel to the motion and perpendicular to it. There is no movement in the direction of the perpendicular component, $F \sin \theta$, and so it does no work. All the work done is by the component that is parallel to the displacement, $F \cos \theta$.

Definition

The work done is equal to the force multiplied by the distance through which the force moves, in the direction of the force.

work done, W (J) = F (N) × s (m) × cos θ

Work is measured in joules (J). One joule of work is done whenever a force of 1 newton moves through 1 metre.

Fig 29

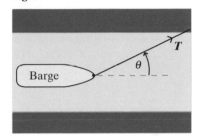

Example

A barge is pulled along a canal by a tow-rope (Fig 29). The tension in the tow-rope is 1000 N and it makes an angle of 20° with the canal. Find the work done in towing the barge 100 m forwards along the canal.

Answer

work done = 1000 × 100 × cos 20° = 94 kJ (to 2 s.f)

Energy and power

Definition

Energy is the ability to do work.

Energy can appear in many different forms. It may be classified as electrical, chemical, kinetic (movement), gravitational potential, electric potential, thermal, electromagnetic (e.g. light), nuclear or sound. Each of these forms of energy can be transferred so that the final effect is doing some work, such as the lifting of a weight. For example, electrical energy can be transferred as kinetic energy by an electric motor. This kinetic energy can be transferred as gravitational potential energy as the axle of the motor winds up a string and lifts a weight. At every stage some energy is transferred as random thermal energy, or heat. In this way, energy tends to become less concentrated and less useful. All forms of energy are measured in the same unit, the joule.

The rate at which a device can transfer energy is known as its **power**. Power is measured as the number of joules per second that are transferred. The S.I. unit of power is the **watt**.

Definition

Power is the rate at which energy is transferred. A power of 1 watt means that 1 joule of energy is transferred every second.

A powerful electric motor can transfer thousands of joules of electrical energy into kinetic energy per second and its power would be given in kilowatts (kW).

Example

An electric shower has a power rating of 7 kW. If the shower runs for 10 minutes, how many joules of energy will have been transferred?

Answer

The shower transfers electrical energy into thermal energy in the water at a rate of 7000 joules per second. In 10 minutes, this is $10 \times 60 \times 7000 = 4\,200\,000\,\text{J}$ or $4.2\,\text{MJ}$

Since energy can result in work being done, power can also be thought of as the rate at which work is done.

Definition

Power, P, is the rate at which work is done, $P = \dfrac{\Delta W}{\Delta t}$

Example

A person of mass 60 kg runs up a flight of stairs that is 10 m high in 6 seconds. Find the power output of the person. (Take $g = 10\,\text{N kg}^{-1}$ to simplify the example.)

Answer

The person has to do work against gravity to lift their own weight.

force $= mg = 60 \times 10 = 600\,\text{N}$

The distance moved in the direction of the force is 10 m. Work done:

$\Delta W = 600\,\text{N} \times 10\,\text{m} = 6000\,\text{J}$

If this work is done in 6 seconds, $\Delta t = 6\,\text{s}$, $P = \dfrac{6000\,\text{J}}{6\,\text{s}} = 1000\,\text{W}$ or $1\,\text{kW}$

Essential Notes

This is the output power of the person. The person would need a greater rate of energy input, since some energy would be transferred as heat. The person is not 100% efficient at transferring chemical energy into work.

For a moving machine, such as a motor or a car, it is often useful to relate the power output to the velocity at which the machine is moving. We can write:

$$\text{power} = \frac{\text{work done}}{\text{time}} = \frac{\text{force} \times \text{distance}}{\text{time}}$$

This can be written as power $= \text{force} \times \dfrac{\text{distance}}{\text{time}}$

Essential Notes

The force, F, and the velocity, v, must be in the same direction. If there was an angle of θ between the direction of the force and the velocity, the equation would become $P = Fv \cos \theta$

Since $\dfrac{\text{distance}}{\text{time}} = \text{velocity}$, power $= \text{force} \times \text{velocity}$:

$$P = Fv$$

Conservation of energy

When energy is transferred from one form to another, the total amount of energy does not change. This idea, known as the conservation of energy, is a fundamental principle in physics. It is not always obvious that the principle is obeyed. When a car brakes to halt at a junction it may appear that all the kinetic energy has disappeared, whereas in fact all the energy has been transferred to the brakes and the surroundings as thermal energy. The conservation of energy only applies to a *closed system*; the total energy will not stay constant if energy has been transferred to another object. In the example of the braking car we have to take into account the energy transferred in heating the air and the road.

Definition

The principle of conservation of energy states that the total energy of a closed system is constant.

Kinetic energy

The energy that a moving mass has because of its motion is known as its **kinetic energy**. The kinetic energy of a moving mass depends on the mass and on the velocity squared.

Definition

Kinetic energy, $E_k = \frac{1}{2}mv^2$

The kinetic energy depends on velocity *squared*, rather than just velocity. So if a car doubles its speed, its kinetic energy will go up by a factor of four. This is why the stopping distance for a car goes up from 6 m at 20 mph to 24 m at 40 mph.

Example

Estimate the average braking force needed if a family car is to be stopped in 6 metres from a speed of 20 mph.

Answer

The mass of a typical family car is just over a tonne, say 1200 kg. Since there are 1.6 km in a mile, a speed of 20 mph is

$$\frac{1600 \times 20}{60 \times 60} = 8.9\,\text{m s}^{-1}$$

The kinetic energy would be:

$E_k = \frac{1}{2}mv^2 = 0.5 \times 1200 \times (8.9)^2 = 47\,400\,\text{J}$

This energy is used to do work against an average braking force F, so $E_k = F \times d$ or

$$F = \frac{E_k}{d} = \frac{47\,400}{6} = 7900\,\text{N}$$

Gravitational potential energy

Gravitational potential energy is the energy that an object has because of its position in a gravitational field. You need to do work to raise a mass to a greater height above the surface of the Earth, and this work is stored as

Essential Notes

Since Einstein's work on relativity, the conservation of energy has had to be extended to include mass. In unit 1 you will have seen that energy can become mass in the process called pair production, where two particles are created from energy. Mass can also become energy in the process of annihilation when matter and antimatter meet and are converted to gamma radiation.

Examiners' Notes

A common mistake is to square the whole expression rather than just the velocity, e.g.

$$\left(\frac{1}{2} \times 1200 \times 8.9\right)^2$$

which would give 2.85 MJ, far too big an answer. Another common mistake is to forget to square the velocity.

potential energy. If the mass is allowed to fall, the potential energy will be transferred to the mass as kinetic energy.

The work done in lifting a mass, m, through a height, Δh, is:

$$\Delta W = \text{force} \times \text{distance} = (m \times g) \times \Delta h$$

This is also equal to the potential energy gained, ΔE_p, by the mass.

> **Definition**
>
> *The change in the potential energy is the mass $\times$ gravitational field strength $\times$ the change in height,*
>
> $$\Delta E_p = mg\Delta h$$

The equation is only strictly true for places where the gravitational field strength, g, is constant. Although g does decrease as the distance from the Earth's surface increases, the equation is reasonably accurate for small values of Δh.

When a mass falls from a height, its potential energy is transferred to kinetic energy. If we can ignore energy losses due to air resistance, then all the potential energy will end up as kinetic energy: $\Delta E_p = \Delta E_k$.

> **Example**
>
> A high diving board is 10 m above the water surface. Calculate the speed at which a diver hits the water.
>
> **Answer**
>
> $$\Delta E_p = \Delta E_k \qquad mg\,\Delta h = \tfrac{1}{2}mv^2$$
>
> So $v^2 = 2g\Delta h = 2 \times 9.81\,\text{N}\,\text{kg}^{-1} \times 10\,\text{m} = 196\,\text{m}^2\,\text{s}^{-2}$
>
> $v = \sqrt{196} = 14\,\text{m}\,\text{s}^{-1}$

Efficiency

During an energy transfer, the total energy stays constant. However, the energy may not all be transferred as *useful* energy. For example, the engine of a car transfers chemical energy from the fuel to kinetic energy, but a significant amount of energy is transferred to thermal energy in the engine, tyres, road surface, etc. The proportion of the input energy that is transferred to useful energy is known as the **efficiency**.

Essential Notes

Efficiency can never be greater than 1. In practice it is always less. It is often expressed as a percentage. For example, the efficiency of a diesel engine is typically between 30% and 40%.

> **Definition**
>
> $$\text{Efficiency} = \frac{\text{useful energy output}}{\text{total energy input}}$$
>
> *Because energy = power $\times$ time, this can also be written*
>
> $$\text{Efficiency} = \frac{\text{useful output power}}{\text{input power}}$$

Example

An incandescent electric light bulb has an input power of 100 W but is only approximately 3% efficient. How much light energy does the bulb emit in one minute?

Answer

$$\text{Efficiency} = \frac{\text{useful output power}}{\text{input power}}$$

So

useful output power = input power × efficiency

In this case the useful power output is in the form of light and is equal to

100 W × 0.03 = 3 W or 3 joules per second

The total light energy output in one minute = 3 × 60 = 180 J

3.2.2 Materials

Bulk properties of solids

Density

The density, ρ, of a material is the mass, m, of a given volume, V.

$$\rho = \frac{m}{V}$$

In S.I. units that means the mass in kg of 1 m^3 of the material. Density is measured in $kg\,m^{-3}$.

To get an idea of the densities of some materials, have a look at Table 4. This section of the specification is about solids, but for comparison the density of water is 1000 $kg\,m^{-3}$ and the density of air is 1.297 $kg\,m^{-3}$ (at 0 °C and standard atmospheric pressure).

Essential Notes

Density is sometimes given in $g\,cm^{-3}$ rather than $kg\,m^{-3}$. To convert $g\,cm^{-3}$ to $kg\,m^{-3}$, you first divide by 1000 to put grams into kilograms, and then multiply by 1 000 000, since there are 1 000 000 cm^3 in 1 m^3. The overall effect is to multiply by 1000, so that water has a density of 1 $g\,cm^{-3}$ or 1000 $kg\,m^{-3}$.

Material	Density/$kg\,m^{-3}$
aluminium	2700
bone	1700−2000
iron	7870
copper	8960
gold	17 650

Table 4
Densities of some common solids

Example

What is the weight of water in a rectangular water tank that measures 0.8 m × 0.8 m × 0.5 m?

Answer

The volume of water is $0.32\,\text{m}^3$. The density of water is $1000\,\text{kg}\,\text{m}^{-3}$.

Since $\rho = m / V$,

$$m = \rho \times V$$

So the mass of water is 320 kg. Taking g to be $9.8\,\text{N}\,\text{kg}^{-1}$, the weight of water $= mg = 3140\,\text{N}$ (to 3 s.f.).

Materials under tensile forces

When an object is subjected to opposing forces, it may be stretched or compressed (see Fig 30).

Fig 30
Tensile (stretching) forces and compressive (squashing) forces

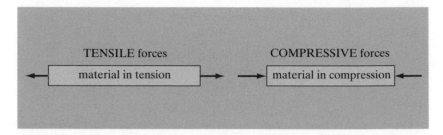

This section of the specification concentrates on the properties of materials under **tensile forces**. Tensile forces tend to stretch the object, putting the material under tensile **stress** and causing an extension in the object.

To investigate the behaviour of a spring under tension, the spring can be suspended vertically. Masses are then hung from the bottom of the spring and the extension is measured (Fig 31). Typical results from such an experiment are shown in Fig 32.

The results show that the extension of a spring is proportional to the force applied to it. This is known as **Hooke's Law** and can be written

$$F = k\,\Delta l$$

where Δl is the change in length of the spring (the extension), and k is the **spring constant**. The spring constant is a measure of the stiffness of the

Fig 31
Stretching a spring

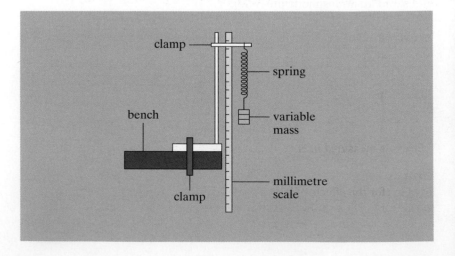

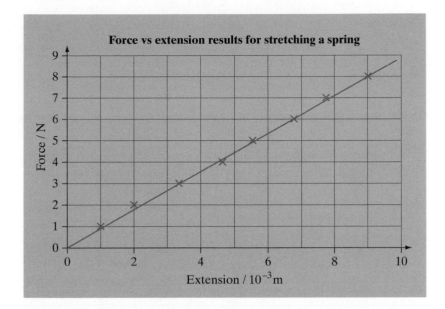

Fig 32
Results from a spring-stretching experiment

spring, i.e. how much force it takes to stretch it by a given distance. Its value is the gradient of the force−extension graph and its units are $N\,m^{-1}$.

Hooke's Law also applies to objects other than springs, for example wires. The law is not obeyed, however, if the force becomes too large − the object is then stretched beyond its limit of proportionality.

Springs (and wires) that obey Hooke's Law are said to show **elastic behaviour**. They return to their original length when the tensile force is removed. However, if the force is large enough it may cause a permanent extension, so that the spring does not return to its original size, even when the force is removed. The spring is said to have been stretched past its **elastic limit**. Objects that show a permanent deformation, even when the force is removed, are said to show **plastic behaviour**.

In order to compare the elastic properties of two different materials, say copper and brass, it is important to take into account the dimensions of the sample. If the metals were in the form of wires for example, the cross-sectional area and the original length would affect how much they stretched for a given force (see Fig 33 overleaf).

To allow for the effects of thicker wires, we define the property **stress** (or tensile stress), σ, as force per unit cross-sectional-area, A.

Definition

$$\text{Stress} = \frac{force}{cross\text{-}sectional\ area} \qquad \sigma = \frac{F}{A}$$

Stress is measured in $N\,m^{-2}$, which is known as the **pascal**, **Pa**.

To allow for the effects of longer wires, we define the property **strain** (or tensile strain), ε, as the fractional extension, that is the extension, Δl, divided by the original length, l.

> **Definition**
>
> $$Strain = \frac{extension}{original\,length} \qquad \varepsilon = \frac{\Delta l}{l}$$
>
> *Because stress is a ratio of two lengths it has no unit.*

Fig 33

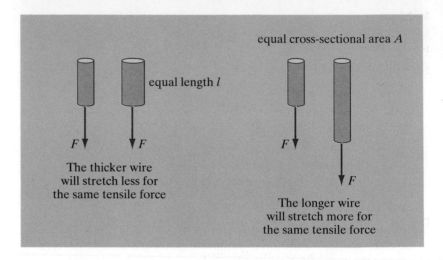

equal cross-sectional area A

equal length l

F F

The thicker wire
will stretch less for
the same tensile force

F

F

The longer wire
will stretch more for
the same tensile force

Different materials behave very differently under tensile stress. For example, imagine trying to stretch a metal wire, a glass fibre or a strip of polythene. We use a number of specialist terms to describe the behaviour of materials under tensile stress:

Elasticity An elastic material returns to its original size and shape when the force which is stretching or compressing it is removed.

Elastic limit This is the maximum stress that can be applied to a material without causing a permanent extension.

Hooke's Law The extension produced by a force in a wire or spring is directly proportional to the force applied. This only applies up to the limit of proportionality.

Yield point Beyond the elastic limit, a point is reached at which there is a noticeably larger permanent change in length. This results in plastic behaviour.

Plasticity A plastic material is the opposite of an elastic material. A plastic material does not return to its original size and shape when the force which is stretching or compressing it is removed. There is permanent stretching and change of shape.

Tensile stress This is the force per unit area of cross section when a material is stretched (units $N\,m^{-2}$ or Pa).

Tensile strain	This is the ratio of extension (Δl) to original length (l). It is the fractional change in length. It has no units.
Strength	Some materials can withstand large stresses before they fracture (break). These are strong or high-strength materials.
Breaking stress (or ultimate tensile stress)	This is the maximum stress that can be applied to a material without it breaking.
Stiffness	This is a measure of how difficult it is to change the size or shape of a material.

- Thick steel wire is stiffer than thin steel wire of the same length.

- Short steel wire is stiffer than longer steel wire of the same diameter.

- Steel is stiffer than copper of the same diameter and length, because the copper extends more per unit force.

Ductility	A ductile material can be easily and permanently stretched. Copper is a good example, and can easily be drawn out into thin wire.
Brittleness	A brittle material cannot be permanently stretched. When a tensile force is applied the material breaks soon after the elastic limit is reached. Brittle materials may be strong in compression. Widespread use of brittle materials, such as cast iron, concrete and house bricks is only possible if the design keeps them in compression.

Materials can be tested by subjecting them to an increasing force, in order to stretch them. The results can be expressed in the form of a force–extension graph (Fig 34) or a stress–strain graph (Fig 35).

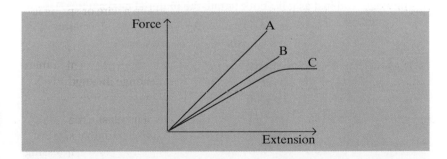

Fig 34
Force–extension graphs for three specimens

Essential Notes

Force–extension graphs apply only to the *specimen* under test. Stress–strain graphs on the other hand apply to the *material* under test, regardless of the dimensions of the specimen.

Specimens A and B in Fig 34 both obey Hooke's Law, since the graphs are straight lines through the origin. However the graph of specimen A has a steeper gradient and so A is stiffer than specimen B. Neither undergoes plastic deformation before breaking, and therefore both are formed from brittle materials. Specimen C obeys Hooke's Law up to the elastic limit, after which it undergoes plastic deformation and is likely to be ductile.

Fig 35
Stress–strain graph

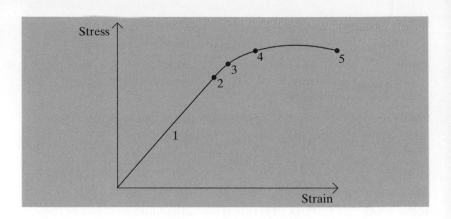

In Fig 35 the following regions and points are marked:

1   Linear region – extension $\propto$ force (obeys Hooke's Law).

2   Proportionality limit – beyond this point extension is no longer proportional to force (limit of Hooke's Law).

3   Elastic limit – material begins to behave plastically. This is the point beyond which, when the stress is removed, the material does not return to its original length.

4   Yield point – material shows large increase in strain for small increase in stress.

5   Breaking stress (or ultimate tensile stress) – the applied stress causes the material to fracture (break).

The behaviour of ductile and brittle materials is different when they are stretched to breaking point.

Ductile materials

When a ductile material (e.g. copper wire) is subjected to a high tensile stress it undergoes considerable plastic deformation (Fig 36). During this plastic stage, and just as it is about to fail, the material **necks** (Fig 37).

Fig 36
Stress–strain graph of a ductile material

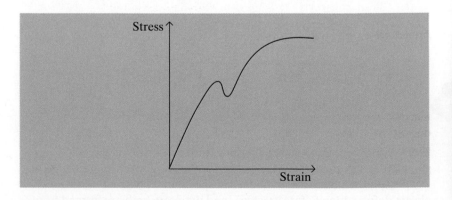

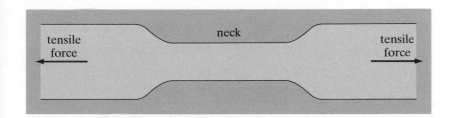

Fig 37
Necking of a ductile material
under stress

The cross-sectional area of the wire gets narrower, and because stress = force/area, the neck experiences an increase in tensile stress. The material fails at the neck. The neck provides an early warning that the material is about to fail.

Brittle materials

Brittle fracture is due to the rapid extension of surface or internal cracks in the material. The fracture is sudden and catastrophic. There is little or no plastic deformation as a warning (Fig 38). There is a concentration of stress around the cracks; the sharper the crack, the greater the stress. Crack movement is easier under tension and more difficult under compression. Pre-stressing brittle materials can increase their strength.

Elastic strain energy

When a wire is stretched by a force, provided the elastic limit is not exceeded, then the work done (energy change) is stored as elastic potential energy, or **elastic strain energy**, in the wire. The area below the graph line is the total work done in stretching the wire, and is therefore equal to the elastic strain energy stored. This can be shown as follows.

Word done by force F in extending the wire by a small extension δl is

$$\delta W = F\,\delta l$$

Total work done in fully extending the wire is therefore the sum of the small areas $F\,\delta l$ (see Fig 39, overleaf).

$$W = \sum \delta W = \sum F\,\delta l$$
$$= \text{area under graph}$$

Area of triangle $= \frac{1}{2} \times \text{base} \times \text{height} = \frac{1}{2} \times F \times \Delta l$

Elastic strain energy $= \frac{1}{2}F\Delta l$

Elastic strain energy is more obvious in a spring where the energy stored during stretching is released again to restore the spring to its original shape.

Since $F = k\,\Delta l$ for a spring (see page 32)

Elastic strain energy stored in a spring $= \frac{1}{2}k(\Delta l)^2$

Fig 38
Stress–strain graph of a brittle
material

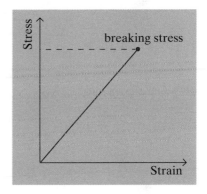

Examiners' Notes

You should be able to present a derivation of the equation energy stored $= \frac{1}{2}F\Delta l$ in an examination.

Fig 39
Calculating the work done in stretching a wire

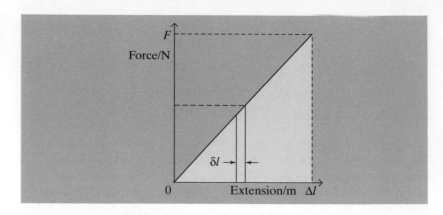

The Young modulus

Definition

*The **Young modulus** (E) is a measure of the stiffness of a material, and is given by:*

$$E = \frac{\text{tensile stress}}{\text{tensile strain}}$$

The Young modulus has units of N m^{-2}, or pascal, Pa.

If F is the tensile force, A is the cross-sectional area, Δl is the extension and l is the original length,

$$E = \frac{F}{A} \div \frac{\Delta l}{l} = \frac{F}{A} \times \frac{l}{\Delta l} \quad \text{or} \quad \frac{Fl}{A\Delta l}$$

The Young modulus can be used to compare the stiffness of different materials, even if the samples under test have different dimensions.

The Young modulus applies only up to the limit of proportionality.

Example

A 1.5 m long steel piano wire, with a diameter of 1 mm, is stretched by a force of 50 N.

(i) Calculate the stress in the wire.

(ii) Calculate the increase in length of the wire.

(iii) Calculate the energy stored in the wire.

(Young modulus of steel = 210×10^9 Pa)

Answer

(i) Cross-sectional area of wire $= \pi\left(\dfrac{d}{2}\right)^2 = \pi \times (0.5 \times 10^{-3})^2$

$$= 7.85 \times 10^{-7} \, \text{m}^2$$

$$\text{Stress} = \frac{\text{force}}{\text{area}} = \frac{50}{7.85 \times 10^{-7}} \, \text{N m}^{-2} = 6.37 \times 10^7 \, \text{Pa}$$

(ii) Strain $= \dfrac{\text{stress}}{E} = \dfrac{6.37 \times 10^7}{210 \times 10^9} = 3.03 \times 10^{-4}$ (no units)

Extension $\Delta l = l \times$ strain $= 1.5 \times 3.03 \times 10^{-4} = 4.55 \times 10^{-4}\,\text{m}$

(iii) Energy $= \frac{1}{2} F \Delta l = 0.5 \times 50 \times 4.55 \times 10^{-4} = 1.1 \times 10^{-2}\,\text{J}$

Experimental determination of the Young modulus of a material

The Young modulus of a material in the form of a wire can be determined using the apparatus shown in Fig 40.

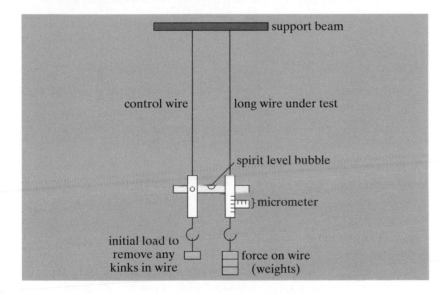

Fig 40
Experimental determination of the Young modulus

Essential Notes

This is often referred to as Searle's apparatus.

A long thin wire (at least one metre) is stretched by loading it with weights and the extension is measured at each loading. The two variables are plotted to produce a straight line graph through the origin, the gradient of which is used to calculate the Young modulus. The key points of the experimental method are:

- two identical wires are fixed in parallel

- both wires are initially loaded to straighten them

- the micrometer attached to the test wire is adjusted, in order to bring the spirit level horizontal

- the micrometer reading is then taken

- a metre rule is used to measure the original length of the test wire, l

- a second micrometer is used to measure the diameter of the wire in several places to account for any unevenness in the wire diameter, and the average reading is used to calculate the cross-sectional area, A, of the wire

- the test wire is then loaded, and the micrometer attached to the test wire is adjusted to bring the spirit level horizontal once more

- the extension is then calculated

Fig 41
Force–extension graph for a wire under test

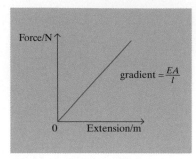

The graph is a straight line since:

$$y = mx + c$$

$$F = \left(\frac{EA}{l}\right)\Delta l + 0$$

The Young modulus can be calculated from the force–extension graph:

$$E = \text{gradient} \times \frac{l}{A}$$

The following measures improve the accuracy of the experiment:

- a long thin wire is used to produce as large an extension as possible for every load added

- a control wire is used so that changes in length due to temperature changes, or to a sagging support, do not affect the results

- a metre rule is accurate enough for measuring the length of the wire since an error of a few mm in 1 m is not significant.

$$\left(\text{e.g. } \frac{2}{1000} \times 100 = 0.2\%\right)$$

- a micrometer with a resolution of 0.01 mm is needed to measure the diameter and extension of the wire. For a wire of 0.5 mm diameter, this gives an error of 0.01/0.5 × 100 = 2% in the diameter. This is a *significant* error, because it leads to an error of 2 × 2 = 4% in the cross-sectional area, A.

- further loads are added, and the procedure repeated until a range of readings has been obtained

- a second set of results is obtained during unloading

- a graph of force against extension is plotted (Fig 41).

Essential Notes

The gradient of a stress–strain graph is equal to the Young modulus of the material.

Example

A material in the form of a wire, 1.0 m long, is subjected to a tensile force and a stress–strain graph of loading and unloading the wire is drawn. The graph is shown in Fig 42.

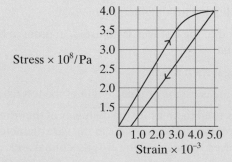

Fig 42

(i) Explain the shape of the graph.

(ii) Use the graph to determine at what extension the limit of proportionality occurs.

(iii) Use an appropriate part of the graph to determine the Young modulus for the material.

(iv) Use the information in Table 5 to determine which material the wire is made of.

Material	Young modulus/Pa $\times 10^9$
aluminium	71
brass	100
copper	117
gold	71
iron	206
silver	70
stainless steel	200
zinc	110

Table 5

Answer

(i) For stresses of up to 3.5×10^8 Pa (the limit of proportionality) the wire obeys Hooke's Law (extension $\propto$ force). Beyond this point the elastic limit/yield point is reached. The material undergoes plastic deformation. This indicates that the material is ductile. When the stress is removed the material has undergone a permanent change in length. The graph does not return through the origin.

(ii) Strain $= \dfrac{\Delta l}{l}$ so extension $\Delta l =$ strain $\times l = 3.0 \times 10^{-3} \times 1.0$

$= 3.0 \times 10^{-3}$ m or 3 mm

(iii) $E = \dfrac{\text{stress}}{\text{strain}} = \dfrac{3.5 \times 10^8}{3.0 \times 10^{-3}} = 1.17 \times 10^{11}$ Pa

(iv) Table 5 reveals that copper has a Young modulus of 117×10^9 Pa.

Progressive waves

When a stretched spring is vibrated at one end, the oscillations are transferred along the spring in the form of a travelling or **progressive wave**.

Fig 43
A progressive wave on a spring

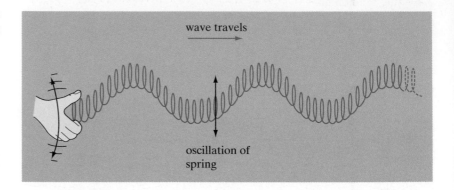

The progressive wave transfers energy from one end of the spring to the other, but the coils of the spring are not permanently displaced by the wave. As the wave passes the coils simply oscillate from side to side before returning to their equilibrium positions.

All mechanical waves travel in the same way – repeated oscillations of the particles are transferred through the medium. A sound wave travels through air as the air molecules vibrate backwards and forwards; a ripple travels across the surface of a pond as the water molecules undergo oscillations. The **amplitude**, A, of a wave is the maximum displacement from equilibrium caused by the wave (see Fig 45).

Wave speed, frequency and wavelength

The speed of a mechanical wave depends on the properties of the medium that it is travelling through. In particular it depends on the strength of the forces between adjacent particles, that is the **elasticity** of the medium, and on the **inertia** of the medium, that is the resistance to acceleration. Wave speed, c, is measured in metres per second, m s^{-1}.

Fig 44
A mechanical wave

(a) You can think of the particles of a medium as masses connected together by springs

(b) Large masses are hard to accelerate and so the wave will travel slowly through the medium. Stiff springs (those with a high value of the spring constant k) will cause large forces and the wave will travel quickly

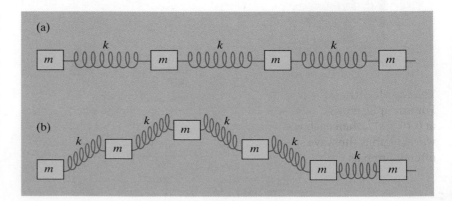

The **frequency** of a wave is determined by the frequency of the oscillations that caused it. If you dip your finger into a pond five times per second, there will be five waves per second spreading out from your finger.

Definition

The frequency of a wave is the number of cycles that occur in one second. The unit of frequency is the hertz, Hz. A frequency of 1 Hz is one wave per second.

The distance between identical consecutive points on a wave is known as the **wavelength**, λ, which is measured in metres.

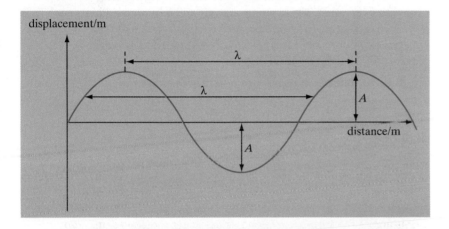

Fig 45
Wavelength and amplitude

The wavelength, λ, of a wave is the distance between any two identical consecutive points on the wave

The amplitude, A, of a wave is the maximum displacement from equilibrium caused by the wave

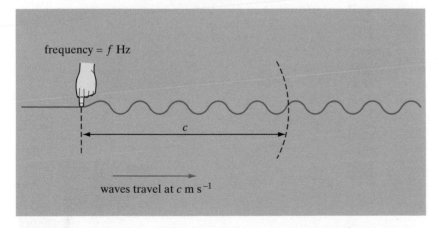

Fig 46
Wave speed, frequency and wavelength

In one second the first wave has travelled a distance of $c\,\text{m}\,\text{s}^{-1}$ and there are f waves. The distance occupied by each wave, λ, is therefore $\lambda = c/f$

The wavelength is determined by the wave speed and the frequency. For example, if there are five water waves per second and the waves travel at $0.10\,\text{m}\,\text{s}^{-1}$, then each wave must occupy $0.10/5 = 0.02\,\text{m}$. In general, the wavelength is the wave speed divided by the frequency, $\lambda = c/f$. This is often written as

$$c = f\lambda$$

Essential Notes

The frequency of electromagnetic waves can be very high. Make sure that you know the SI prefixes:

kHz = kilohertz = 10^3
MHz = megahertz = 10^6
GHz = gigahertz = 10^9
THz = terahertz = 10^{12}

Example

Radio 1 is transmitted on a frequency of 98 MHz. Calculate the wavelength of a radio wave of this frequency. (The speed of electromagnetic waves is $3 \times 10^8\,\mathrm{m\,s^{-1}}$.)

Answer

$$\lambda = \frac{c}{f} = \frac{3 \times 10^8}{98 \times 10^6} = 3.06\,\mathrm{m}$$

Phase and path difference

We can sketch a graph showing the displacement caused by a wave against distance. This graph shows a wave frozen at an instant in time. Points that are a whole wavelength apart, such as A and B in Fig 47, are oscillating in time with each other. These points are said to be **in phase**. Points such as B and C are oscillating **in antiphase** with each other. The difference in phase between two points on the same wave is expressed as an angle, ϕ. This may be either in radians or in degrees.

The **phase difference** between two points at distances x_1 and x_2 is

$$\phi = 2\pi\left(\frac{x_1 - x_2}{\lambda}\right) \text{ in radians} \quad \text{or} \quad \phi = 360\left(\frac{x_1 - x_2}{\lambda}\right) \text{ in degrees}$$

If the distance between x_1 and x_2 is a whole number of wavelengths then $\phi = 2\pi, 4\pi, 6\pi$, etc, in radians, or multiples of $360°$.

Fig 47

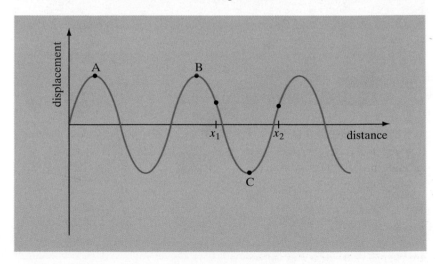

Fig 48
The path difference between the two waves is $S_2P - S_1P$

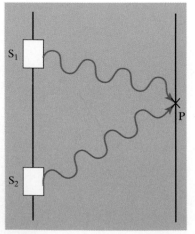

Suppose that there are two wave sources, S_1 and S_2 (Fig 48), emitting waves of the same wavelength that are in phase on leaving the sources. When they arrive at a point P their relative phase depends on the distance travelled by each wave. The difference in the distance travelled by the two waves, $S_2P - S_1P$, is called the **path difference**. Then, at P,

$$\text{phase difference} = 2\pi \times \left(\frac{\text{path difference}}{\lambda}\right) \text{ in radians}$$

Longitudinal and transverse waves

Waves can be classified into two groups according to the direction of their oscillations.

Longitudinal waves

Longitudinal waves have vibrations that are parallel to the direction in which the wave is travelling (Fig 49). Sound waves are longitudinal waves, since the air molecules vibrate backwards and forwards in a direction parallel to the wave's travel (Fig 50). Seismic P-waves and compression waves in a spring are also examples of longitudinal waves.

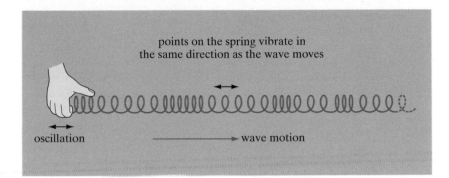

Fig 49
A progressive longitudinal wave on a spring

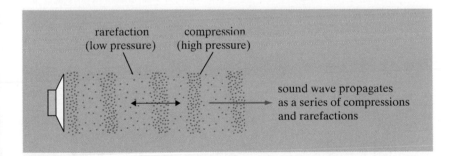

Fig 50
Sound travels as a longitudinal wave

As a loudspeaker vibrates backwards and forwards the molecules of air are pushed closer together, as a **compression**, and then further apart, as a **rarefaction**. Compressions are regions of higher pressure and density; rarefactions are regions of lower density and pressure.

Transverse waves

Transverse waves have vibrations that are perpendicular to the direction of propagation. Water waves, waves on strings (Fig 51) and seismic S-waves are examples of transverse waves.

Electromagnetic (EM) waves, like light, are also transverse waves but there is a significant difference between these waves and the other waves discussed so far. EM waves are not vibrations of particles in a medium, but are oscillating electric and magnetic fields (Fig 52).

These fields are perpendicular to each other and to the direction of wave travel. EM waves do not need a medium to support the wave, indeed they travel fastest in a vacuum, at a speed of

$$c = 3 \times 10^8 \, \text{m s}^{-1}$$

Fig 51
Transverse wave, for example on a string

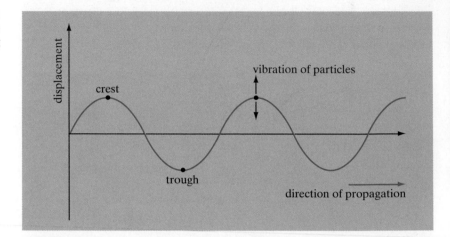

Fig 52
Electromagnetic wave

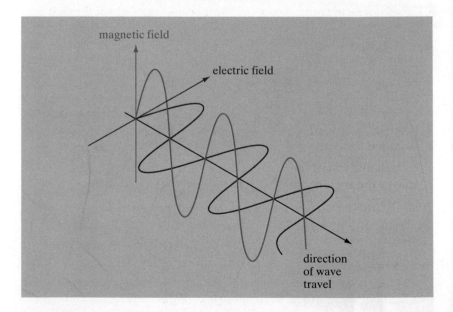

Polarisation

A transverse wave has oscillations in a plane that is perpendicular to the wave's velocity. However, these oscillations could be in any direction in that plane (Fig 53). The electric field in a light wave, for example, can oscillate in any direction in that plane. Such a wave is said to be unpolarised.

It is possible to restrict the oscillations of the electric field in an EM wave to one direction only. The wave is then said to be **polarised**.

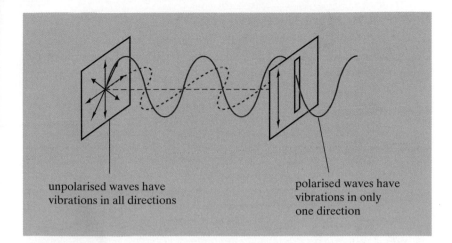

Fig 53
Polarisation

The polarisation of electromagnetic waves has important applications.

- The radio waves that are used to carry TV signals are transmitted as horizontally or vertically polarised waves. Your TV aerial must be aligned in the same plane. Neighbouring transmitters may emit waves of opposite polarisations. This reduces interference between the two signals.

- It is possible to produce polarised light by passing the light through a sheet of Polaroid. Two sheets of Polaroid at right angles to each other will effectively block off all the light. Since reflected light is partly polarised, Polaroid sunglasses can reduce the glare from reflected light.

- Some materials, like sugar solution, can twist the direction of polarisation of light. These materials are said to be 'optically active'. Perspex is optically active, and the angle through which the direction of polarisation is twisted depends on the stress the material is under. Polarised light can therefore be used to examine stress patterns (Fig 54).

- Some liquid crystals are optically active. An applied electric field can change the angle of polarisation. This effect is the basis of liquid crystal displays, LCDs.

Essential Notes

The fact that light can be polarised is a crucial piece of evidence that suggests that light travels as a transverse wave.

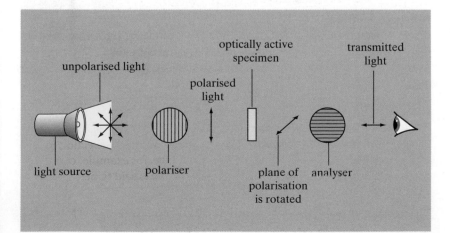

Fig 54
Arrangement for viewing stress patterns

Refraction at a plane surface

When a wave moves from one medium into another, such as a light wave moving from air into glass, the wave changes speed. This can lead to the wave changing direction.

> **Definition**
>
> *The change of direction when a wave moves from one medium into another is called **refraction**.*

Fig 55
Light waves travelling at right angles into a pane of glass, showing the change of wavelength (exaggerated). Here there is no change of direction, but the wavelength and speed of the waves are reduced

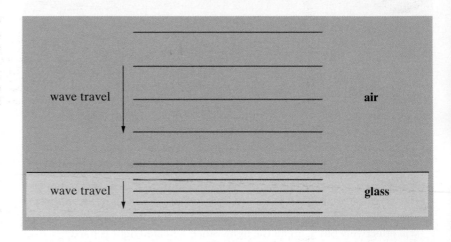

Essential Notes

When light waves pass through glass, the frequency of the waves must stay the same. After all if 5×10^{14} waves per second enter the glass then the same number must come out – if not, where would they go?

The speed of light in air is almost $3 \times 10^8 \, \text{m s}^{-1}$, but in glass it is less, about $2 \times 10^8 \, \text{m s}^{-1}$. As a light wave travels into a glass window the wavefronts become closer together as the wave slows down (Fig 55). The speed c and the wavelength λ both decrease but the frequency f stays the same.

If the wavefront hits the boundary between the two media at an angle, then the wavefront changes direction (Fig 56). A ray of light refracts towards the normal when it slows down, and away from the normal when it speeds up. The greater the change in speed, the more the ray deviates from its original path.

Refractive index

All electromagnetic waves, including light waves, travel at the same speed c in a vacuum. Precise measurements give this value as $c = 299\,792\,458 \, \text{m s}^{-1}$. As the waves pass through matter they slow down to a new speed v.

Essential Notes

The refractive index of air is 1.0003 but this is usually taken as being equal to 1. In other words we usually assume that light travels at the same speed in air as in a vacuum.

> **Definition**
>
> *The **refractive index** n of a material is the ratio of the speed of light in a vacuum to its speed in the material.*
>
> $$n = \frac{\text{speed of light in a vacuum}}{\text{speed of light in the material}} = \frac{c}{v}$$
>
> *This is sometimes referred to as the absolute refractive index.*

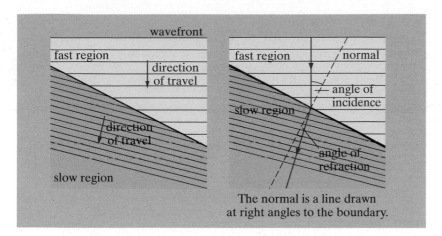

Fig 56
If waves hit a boundary between two media at an angle, the change in speed causes a change in direction

Because light always travels faster in a vacuum than in any other material, the value of the absolute refractive index is always greater than 1. It is a ratio so has no unit.

Refractive indices of some common materials are listed in Table 6. The higher the value of the refractive index, the more the light is slowed down and the greater its deflection. Materials with a high value of refractive index are said to be 'optically dense'.

When a light wave passes from one material into another, say from water into glass, it is the relative speed of light in each of the materials that determines how much a ray will be refracted. The **relative refractive index** is

Table 6
Refractive indices

Diamond	2.42
Glass	1.5 to 2.0
Perspex	1.50
Water	1.33
Sea water	1.34
Ice	1.31

$$_1n_2 = \frac{\text{speed of light in medium 1}}{\text{speed of light in medium 2}} = \frac{v_1}{v_2}$$

If we know the absolute refractive indices for two materials, say n_1 and n_2, it is possible to calculate the relative refractive index for a light wave moving between them:

$$_1n_2 = \frac{v_1}{v_2} = \frac{c}{v_2} \times \frac{v_1}{c} = n_2 \times \frac{1}{n_1}$$

So

$$_1n_2 = \frac{n_2}{n_1}$$

The relative refractive index can have a value of less than 1. For example, the relative refractive index for a light wave moving from water to ice is

$$_{\text{water}}n_{\text{ice}} = \frac{n_{\text{ice}}}{n_{\text{water}}} = \frac{1.31}{1.33} = 0.985$$

Essential Notes

The relative refractive index for a wave moving from medium 1 to medium 2 is the inverse of the relative refractive index for a wave moving in the opposite direction, from medium 2 to medium 1:

$$_1n_2 = \frac{1}{_2n_1}$$

Snell's Law of refraction

The relative refractive index between two materials affects how much a ray of light will deviate at the boundary. It can be shown (see Fig 58 below) that

$$_1n_2 = \frac{\sin(\text{angle of incidence})}{\sin(\text{angle of refraction})}$$

So, in Fig 57,

$$\frac{n_2}{n_1} = \frac{\sin\theta_1}{\sin\theta_2}$$

or

$$n_1\sin\theta_1 = n_2\sin\theta_2$$

This is **Snell's Law**.

Fig 57

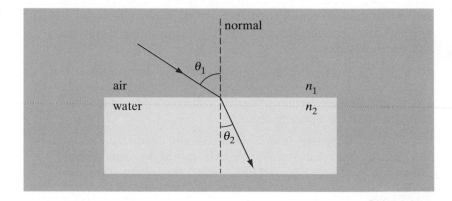

Essential Notes

Another way to put this is that a ray of light will refract *towards* the normal as it enters an optically denser medium, and *away* from the normal as it leaves again.

If the light waves are travelling into an optically denser material, say from air into water, they will slow down, and θ_2 will be less than θ_1. The angle is always smaller in the optically denser material.

Fig 58
Deriving Snell's Law

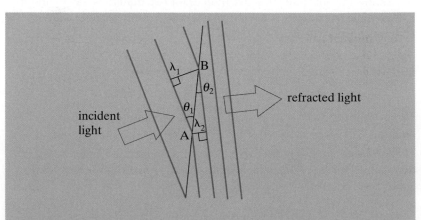

Essential Notes

Snell's law applies (a) when a light ray passes from one medium to another, (b) for monochromatic light (single wavelength) and (c) to rays which lie in the same plane.

Adjacent wavefronts are separated by one wavelength. As the wave crosses the boundary, its frequency remains constant so the change of speed causes a change of wavelength, from λ_1 to λ_2. By definition, $v = f\lambda$ and

$$_1n_2 = \frac{n_2}{n_1} = \frac{v_1}{v_2} \text{ so } \frac{n_2}{n_1} = \frac{f\lambda_1}{f\lambda_2} = \frac{\lambda_1}{\lambda_2}$$

Using $\lambda_1 = AB \sin\theta_1$ and $\lambda_2 = AB \sin\theta_2$,

$$\frac{n_2}{n_1} = \frac{AB \sin\theta_1}{AB \sin\theta_2} = \frac{\sin\theta_1}{\sin\theta_2}$$

It is easy to show that θ_1 is equal to the angle of incidence and θ_2 is equal to the angle of refraction.

Example

A ray of light strikes a glass block at an angle of 30° to the normal. Find the angle of refraction in the glass block. (Take the speed of light in glass to be $2.0 \times 10^8 \, \mathrm{m\,s^{-1}}$ and the speed of light in air to be $3.0 \times 10^8 \, \mathrm{m\,s^{-1}}$.)

Answer

Relative refractive index $_1n_2 = \dfrac{3.0 \times 10^8 \, \mathrm{m\,s^{-1}}}{2.0 \times 10^8 \, \mathrm{m\,s^{-1}}} = 1.5$

$$_1n_2 = \frac{\sin\theta_1}{\sin\theta_2}$$

and $\theta_1 = 30°$

So

$$\sin\theta_2 = \frac{\sin 30°}{1.5} = 0.33$$

Therefore

$\theta_2 = 19.5°$

Total internal reflection

Definition

Total internal reflection is when a ray of light, leaving an optically dense material and travelling into a less dense one, is not refracted out of the dense material but is totally reflected back inside.

When light is travelling from glass into air, for example, if the angle of incidence in the glass is high enough the light is no longer refracted out of the glass. This is total internal reflection.

We can use Snell's Law to calculate the angle of refraction in each of the cases shown in Fig 59.

Fig 59
Refraction and internal reflection

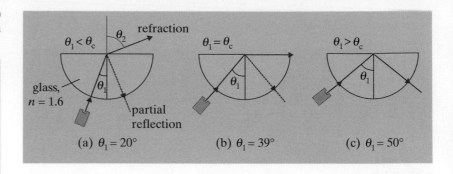

(a) $\theta_1 = 20°$ (b) $\theta_1 = 39°$ (c) $\theta_1 = 50°$

Examiners' Notes

Total internal reflection occurs when the light ray is moving from one medium to another in which the speed of light is greater, *and* the angle of incidence is greater than the critical angle.

If the refractive index from air to glass is 1.6, then the refractive index from glass to air is

$$\frac{1}{1.6} = 0.625$$

(a) $\sin\theta_2 = \dfrac{\sin\theta_1}{n} = \dfrac{\sin 20°}{0.625} = 0.547$

so the angle of refraction is 33.2°.

(b) $\sin\theta_2 = \dfrac{\sin\theta_1}{n} = \dfrac{\sin 39°}{0.625} = 1.00$

so the angle of refraction is 90°. This means that the refracted ray is deviated so much that it just grazes the surface of the block.

(c) $\sin\theta_2 = \dfrac{\sin\theta_1}{n} = \dfrac{\sin 50°}{0.625} = 1.23$

It is impossible for the sine of an angle to be greater than 1. Snell's Law is not applicable here because the ray is no longer refracted, but is totally reflected inside the block.

The change from refraction to reflection is not sudden; some light is always reflected inside the block. As the angle of incidence increases, more and more of the light is reflected. At one particular angle of incidence, the **critical angle**, the refracted ray disappears (Fig 59(b)). At this angle the ray is trying to travel along the boundary between the two materials. If the angle of incidence is greater than the critical angle then *all* the incident light is reflected and none of it escapes.

The value of the critical angle depends on the refractive indices of the two media. At the critical angle θ_c, the angle of refraction is 90°, so

$$\frac{n_1}{n_2} = \frac{\sin\theta_c}{\sin 90°} = \sin\theta_c$$

Where the rays of light are moving from a medium of refractive index n into air, which has a refractive index of 1, then

$$\sin\theta_c = \frac{1}{n}$$

Example

An optical fibre (see page 54) is made from glass of refractive index 1.5. A ray of light strikes the end of the fibre at an angle of 75° to the axis of the fibre. Sketch the situation, showing the path of the ray through the fibre.

Answer

Since the angle of incidence is 75°, and the refractive index is 1.5, the angle of refraction is given by

$$\sin \theta_2 = \frac{\sin 75°}{1.5}$$

So $\theta_2 = 40°$

The ray will strike the wall of the fibre at an angle of 50°. This is above the critical angle for this glass:

$$\sin \theta_c = \frac{1}{n} = \frac{1}{1.5} = 0.67, \text{ so } \theta_c = 42°$$

This means that the ray is totally reflected inside the glass. The angle of reflection will be equal to the angle of incidence, 50°.

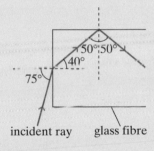

incident ray   glass fibre

A prism can be used to reflect light (Fig 60). The reflection is clearer than with a mirror, where silvering causes multiple reflections. Totally reflecting triangular prisms are used in preference to mirrors in binoculars to invert the image.

It is total internal reflection which makes precious stones sparkle; the refractive index of diamond is very high, around 2.4. Total internal reflection is also responsible for the mirage which makes a tarmac road appear shiny on a hot day.

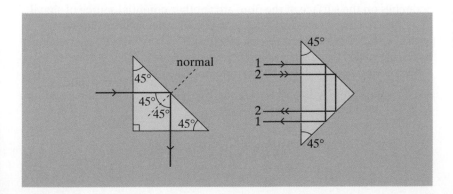

Fig 60
A prism used in two ways to reflect light

Fibre optics

The most important application of total internal reflection is in **optical fibres**. Optical fibres carry cable TV and telephone communications and form the backbone of computer networks.

In its simplest form an optical fibre is a thin strand of pure glass. Light is refracted into one end and strikes the internal wall of the fibre at an angle of incidence greater than the critical angle. Total internal reflection occurs and the ray of light is confined within the fibre (Fig 61).

If the refractive index of the glass fibre is 1.60, then the critical angle is given by

$$\sin\theta_c = \frac{1}{n} = \frac{1}{1.60} = 0.625$$

This gives a critical angle of $\sin^{-1} 0.625$, or 39°. Any rays of light that strike the inner surface of the glass at angles greater than this will be reflected inside the fibre.

Fig 61
General path of light through an optical fibre

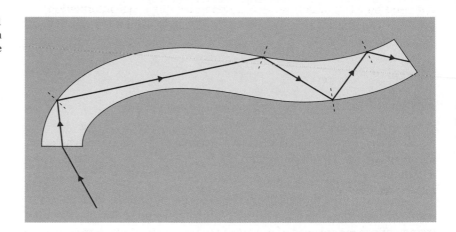

Essential Notes

The number of digital (on-off) pulses that can be transmitted per second along a fibre is known as its bandwidth. Optical fibres offer much greater bandwidth than conventional copper cables.

Optical fibres are used to carry information in the form of digital pulses of infrared light over long distances. It is important that the light losses are as low as possible. This is achieved by using glass with a very low level of impurities, since impurity atoms can scatter the light so that it strikes the boundary at less than the critical angle and is refracted out of the fibre.

Scratches on the surface of the fibre are another potential source of signal loss (Fig 62). One way to avoid this is to use an outer layer of different glass, known as **cladding**, to protect the inner, or core, fibre. The cladding has to be made from a glass of lower refractive index than the core, otherwise total internal reflection could not take place.

For an optical fibre with a core of refractive index $n_1 = 1.60$ and a cladding of $n_2 = 1.50$, the critical angle is given by

$$\sin\theta_c = \frac{n_2}{n_1} = \frac{1.5}{1.6} = 0.9375$$

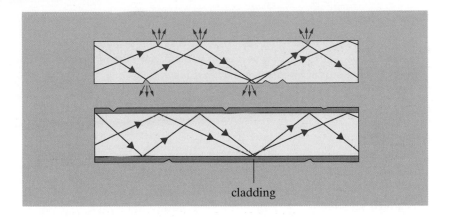

cladding

Fig 62
Surface scratches lead to light leakage in a single fibre. Scratches in the outer cladding do not matter because it does not carry a light signal

So the critical angle is 69.6°.

Optical fibres are also used in medicine to enable doctors to look into the body without the need for invasive surgery. In gastroscopy a bunch of optical fibres is used to carry light down the oesophagus into the stomach. A second bunch of optical fibres is used to carry the image information back to a video camera. Similar devices, collectively known as **endoscopes**, are used to look inside other parts of the body. In some cases the fibres can carry intense laser light, which is used instead of a scalpel as a tool in surgery.

Superposition of waves, stationary waves

When two similar waves meet, the resultant wave depends on the amplitude and the relative phase of the two waves. For example, when two identical water waves meet, perhaps when a wave on the sea reflects from a harbour wall, they may add together to give a double height wave or they may cancel each other out, leading to a region of stillness.

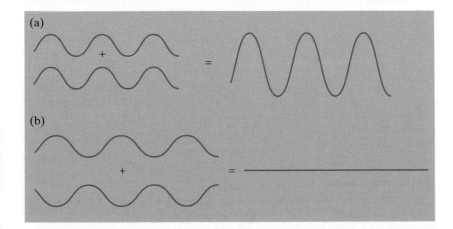

Fig 63
Superposition of waves

(a) Waves arrive in phase and add together constructively

(b) Waves arrive in antiphase and add destructively

Definition

*The **principle of superposition** says that the resultant displacement caused by two waves arriving at a point is the vector sum of the displacements caused by each wave at that instant.*

Example

Two square pulses of wavelength 20 cm are approaching each other on a string. The waves are initially 20 cm apart and they are travelling towards each other at 20 cm s^{-1}. Sketch the string

(a) after 1 second

(b) after 2 seconds.

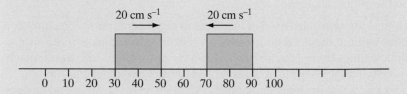

Answer

The waves will overlap after 1 second and the resulting amplitude will be the sum of the individual amplitudes. After 2 seconds the waves have passed through each other.

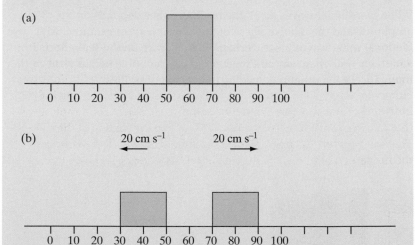

Stationary waves

When two continuous similar waves are travelling in opposite directions, they can superpose to form a **stationary wave** (or standing wave). A stationary wave is a fixed pattern of vibration. Unlike a progressive wave, no energy is transferred along the wave.

On a progressive wave, each point on the wave has the same amplitude. This is not true for a standing wave, where the maximum displacement of any point depends on its position. There are some points on a standing wave which do not vibrate at all. These are called **nodes**. At a node the waves travelling in opposite directions always add together to give zero displacement. The distance between two nodes is always half a wavelength. In between the nodes there are positions of maximum displacement, known as **antinodes**.

All the points between any two nodes vibrate in phase, whereas in a progressive wave the phase changes with position along the wave (see page 44). Points in adjacent loops in a stationary wave vibrate in antiphase.

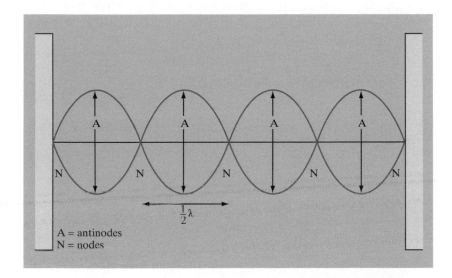

Fig 64
Stationary wave

Stationary waves on the strings of an instrument, such as a guitar or a violin, are the source of vibration for musical notes. When the string is made to vibrate, by being plucked or scraped with a bow, reflections from either end superpose to cause the stationary wave. Since a string is fixed at both ends, there must be a vibration node at each end. The simplest way that a string can vibrate is with one antinode in the middle of the string. This wave pattern is known as the fundamental mode. The frequency of the fundamental mode, the **fundamental frequency** f_0, is given by c/λ.

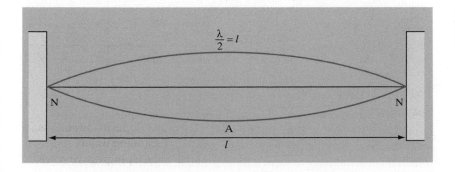

Fig 65
The fundamental mode of vibration for a string

If the string has a length, l, the wavelength is $2l$, since there is half a wave on the string. This gives the fundamental frequency as

$$f_0 = \frac{c}{2l}$$

where c is the speed of the wave along the string.

The string can also support oscillations which have a node in the centre of the string. This is called the first **overtone** (or the second harmonic). There is now a whole wave on the string, so $\lambda = l$. The frequency of the first overtone is

$$f_1 = \frac{c}{l}$$

This is twice the frequency of the fundamental.

Examiners' Notes

Watch out for the difference between harmonics and overtones. The fundamental is the first harmonic.

Fig 66
The overtones for a wave on a string

(a) first overtone (second harmonic)

(b) second overtone (third harmonic)

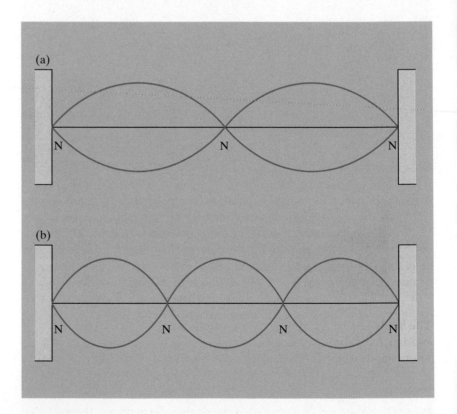

Interference

When two waves meet at a point, by the principle of superposition the combined effect is found by adding together the displacements of the individual waves. If the two waves meet exactly in phase, they will add together constructively to form a double-height wave. If the waves are exactly out of phase they will add together destructively and cancel each other out (see Fig 63).

The principle of superposition applies to all types of waves, including light, and yet we rarely observe two light waves adding together to give darkness. This is because the phase difference between two light waves is always changing. Any point where two waves cancelled each other out would only last for a very short time and would not be detected.

In order to observe a steady **interference pattern** we need to have two waves that maintain a fixed phase difference over a period of time. To achieve this we need two wave sources that emit waves of constant wavelength, which always have the same phase difference. Such wave sources are known as **coherent**.

It is relatively straightforward to generate coherent sound waves. Two identical loudspeakers driven by the same signal generator will produce coherent sound waves.

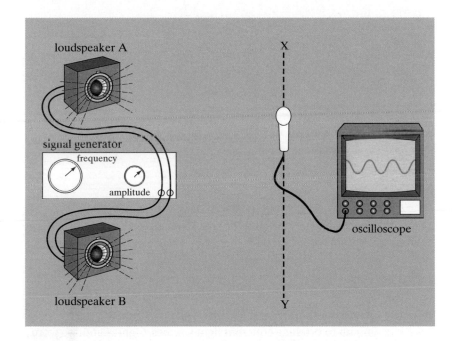

Fig 67
Interference of two sound waves

The loudspeakers A and B (Fig 67) are emitting coherent waves that overlap. A microphone that is moved along the line XY will detect extra-loud regions where the two waves are adding together in phase. There will also be quiet regions, where the two waves arrive out of phase and add together destructively.

When the sound waves leave the speakers, they are in phase. Their relative phase when they arrive at the line XY depends on how far each wave has travelled. If the waves have travelled the same distance (as at point O in Fig 68), they will arrive in phase, and there will be a loud region. In fact, if the path difference between the two waves is any whole number of wavelengths, the waves will add together constructively to give a maximum. A quiet region, or minimum, will occur when the path difference is one half of a wavelength (as at point P in Fig 68).

Fig 68

There is an interference maximum at O, where the path difference between the waves is zero

There is an interference minimum at P, where the path difference between the waves is half a wavelength

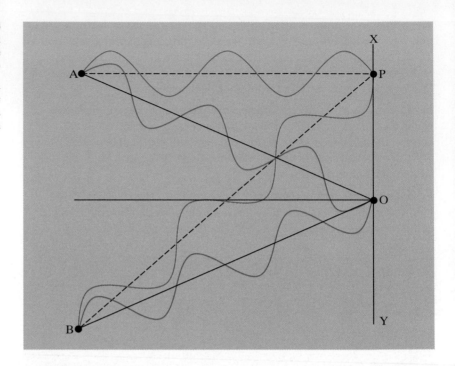

In general, an interference maximum will occur when the path difference is equal to an integral number of wavelengths:

> For a maximum: path difference = $n\lambda$ where $n = 0, 1, 2, 3, \ldots$

A path difference of an odd number of half wavelengths will give rise to an interference minimum:

> For a minimum: path difference = $(2n + 1)\dfrac{\lambda}{2}$ where $n = 0, 1, 2, 3, 4, \ldots$

Two-slit interference patterns (Young's slits)

It is more difficult to produce coherent light sources. Light is emitted from atoms as a result of energy changes by their electrons. These emissions are random and give rise to pulses of light that last for a very short time. These pulses of light (photons) have random phase and random polarisation. Even if we used two **monochromatic** light sources (such as sodium lights), which only emit light of a single frequency, the sources would not have a fixed phase relationship, and would not be coherent. We could not produce a steady interference pattern.

Thomas Young was the first to describe an experiment to produce an interference pattern from light waves. He used a single light source (a candle in fact!) and passed the light through a narrow slit, and then used two slits to divide the wavefront coming from the single slit (Fig 69). The two slits then act as two sources. Since these sources emit light which originally came from the same wave, they must be coherent. Young was able to produce an interference pattern on a screen. The pattern was a series of equally spaced light and dark bands.

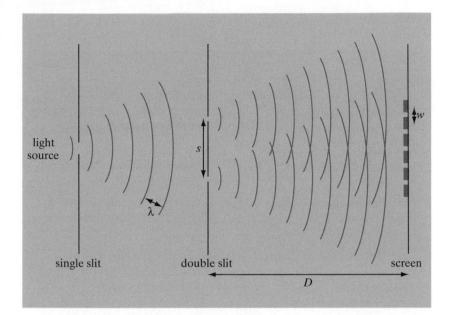

Fig 69
Apparatus for Young's slits
experiment

Essential Notes

Thomas Young's experiment
in 1801 was one of the most
significant pieces of evidence
to support the wave theory
of light. Before Young's
interference experiment, most
scientists accepted Newton's
view that light was a stream of
particles, which Newton called
'corpuscles'. It is impossible to
explain interference patterns
using a particle model of light.

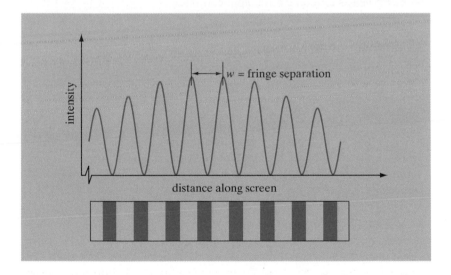

Fig 70
Interference pattern observed in
Young's slits experiment

The light bands (maxima) occur whenever the path difference between
light waves from the two slits is a whole number of wavelengths. The
distance between two successive maxima, *w*, depends on:

- The distance between the two slits, *s*. Increasing *s* makes the
  interference fringes closer together.

- The wavelength of light, λ. The fringes are closer together at short
  wavelengths.

- The distance, *D*, between the slits and the screen. If we observe the
  interference pattern further away from the slits, the fringes are also
  further apart.

The formula that links these quantities is

$$w = \frac{\lambda D}{s}$$

Essential Notes

Laser light can damage eyesight. Because the beam does not diverge, the power is concentrated into a small area which can damage the sensitive cells on the retina of the eye. Scientists using a laser beam for experiments will often use protective goggles. School lasers are not high powered, but should still be treated with respect. You should ensure that the beam cannot accidentally shine into someone's eye. You should also be aware of reflections from the laser. You should use a screen with a matt surface to view the interference patterns.

Although it is possible to produce two-slit interference patterns with an ordinary light bulb and some slits, in practice it is much easier and more effective to use a **laser**. The light from a laser differs in several major respects from that emitted by a light bulb.

- Laser light is monochromatic: all the light is emitted at a single wavelength.

- Laser light is coherent: the light waves emitted by the laser are all in phase, whereas light from an ordinary light bulb is emitted with random phase differences.

- Laser light beams are highly directional: the beam emitted by a laser diverges very little, whereas the light from a light bulb is emitted in all directions.

These characteristics make the laser ideal for demonstrating interference effects, as well as for many other applications.

Example

A laser that emits red light of wavelength 600 nm is used to illuminate two narrow slits 0.1 mm apart. Light from the slits is allowed to fall on a screen 2 m away. Describe and explain what you would see on the screen.

Answer

A laser is a monochromatic source of light. Light that has passed through the two slits will be coherent and will give rise to an interference pattern on the screen. A series of light and dark fringes will be formed. The separation between adjacent bright fringes will be

$$w = \frac{\lambda D}{s} = \frac{600 \times 10^{-9}\,\text{m} \times 2\,\text{m}}{0.1 \times 10^{-3}\,\text{m}} = 0.012\,\text{m}$$

Diffraction

When waves pass through a gap, or travel past an obstacle, the waves spread out. This 'spreading' of waves is known as **diffraction**. Diffraction is a phenomenon common to all types of wave, for example:

- Water waves on the surface of the sea spread out as they pass through the gap in a harbour wall.

- Sound waves diffract through an open doorway.

- Radio waves diffract around hills and buildings.

Since light is a wave motion, it should also diffract and yet we are not usually aware of light spreading round obstacles or through gaps. This is because the amount of diffraction depends upon the relative size of the wavelength and the gap. Diffraction is only noticeable when the wavelength and the gap size are similar.

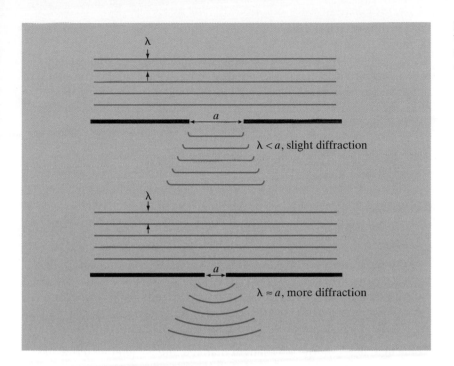

Fig 71
When the wavelength is small
compared to the size of the gap,
diffraction is negligible

Diffraction of light at a single slit

Visible light has a wavelength of between 400 nm and 700 nm. This is small compared to everyday objects and so we do not usually observe diffraction effects. We can observe diffraction of light only if we use a small enough slit.

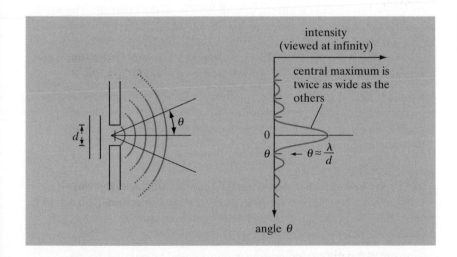

Fig 72
Diffraction of light through
a single slit

The diffraction pattern from a single slit has a series of bright and dark fringes. There is a broad, bright central maximum, with narrower, less bright, secondary maxima on either side. The central maximum gets wider, i.e. there is *more* diffraction, if

- the wavelength is longer
- the gap is smaller.

Light is diffracted as it passes through the pupil of your eye. The amount of diffraction is small, but it could be enough to blur the images of two adjacent objects into one. For example, the light from two adjacent stars may be diffracted so that it is impossible to distinguish them. The effects of diffraction are less when the starlight is observed through a telescope. Since the diameter of the aperture is much greater, there is less diffraction and the images of the two stars can be resolved.

Example

A student looks at two red light bulbs through an adjustable slit. Describe what she would see as the slit is made narrower. What difference would it make if the light bulbs emitted blue light instead?

Answer

As the slit narrowed, diffraction effects would become more noticeable. The light from the two bulbs would begin to spread out and the student would see a central maximum and a set of light and dark fringes from each light bulb. Eventually the two central maxima would overlap and the student would be able to distinguish only one bright object.

Blue light has a shorter wavelength than red and is therefore diffracted less for any particular slit size. The student would notice the same effects as above, but at smaller slit widths.

The diffraction of light through a slit has an effect on the two-slit interference pattern, which was discussed on page 61. Light is diffracted through each slit before it interferes with light from the other slit. The single-slit diffraction pattern is superimposed on the two-slit interference pattern so the intensity of the interference maxima is limited by the diffraction pattern of the single slit. If we use three, four or more slits the interference maxima get sharper (Fig 73). The maxima also get further apart.

If we use hundreds of slits, close together, the interference pattern is just a few sharp bright bands that occur only in certain directions. These directions are determined by the wavelength of the light and by the separation of the slits.

The diffraction grating

A diffraction grating is a series of narrow, parallel slits, usually formed by ruling lines on glass. When light shines on the diffraction grating a set of bright sharp lines are seen. We can derive an expression to calculate the directions in which these maxima will occur.

Assume that the diffraction grating is illuminated by parallel, mono-chromatic light that strikes the grating at right angles (see Fig 74). A and B are identical points in adjacent slits separated by the slit separation, d. The path difference between the light from A and B in the direction θ is therefore $d \sin \theta$. The path difference between all identical points on all the adjacent slits is also $d \sin \theta$. At some angles this path difference will equal a whole number of wavelengths. All the light in these directions will be in phase and there will be an interference maximum.

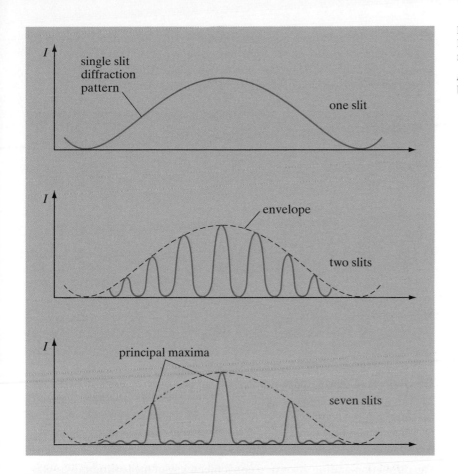

Fig 73
Interference pattern from
multiple slits

As more slits are added the maxima
become sharper and further apart

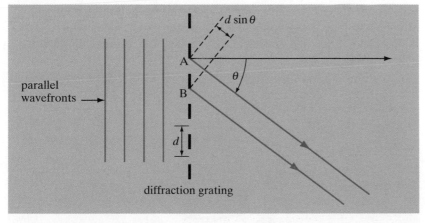

Fig 74
Light incident on a diffraction grating
with uniform slit separation d

For a bright fringe (a maximum) to occur,

$$d \sin \theta = n\lambda$$

where $n = 0, 1, 2, 3, 4, \ldots$.

The first maximum will occur when $n = 0$, i.e. when the light from adjacent
slits has zero path difference. This will happen when $\theta = 0°$. This is the
'straight-through' position and is referred to as the **zero-order maximum**.

Examiners' Notes

You need to know how to
derive this expression.

When $n = 1$ the light from adjacent slits has a path difference of exactly one wavelength, and so all the light waves are in phase. This is the **first-order maximum**. There are two first-order maxima, positioned symmetrically either side of the straight-through beam.

Fig 75
Diffraction maxima from a grating

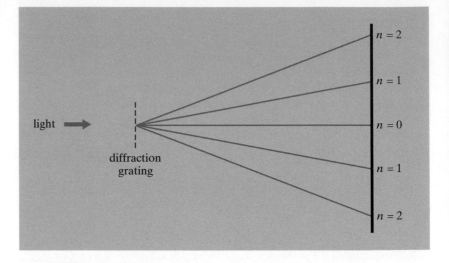

The number of diffraction maxima that can be observed depends on the wavelength and the spacing of the lines on the grating. The order of the maximum, n, is given by $n = (d/\lambda) \sin \theta$. Since the largest possible value for $\sin \theta$ is 1, the highest-order diffraction maximum that can be observed is $n = d/\lambda$.

The diffraction grating gives us a convenient way of measuring the wavelength of light. A device called a **spectrometer** is used (see Fig 76). The spectrometer uses a collimator to produce parallel light, which illuminates a diffraction grating at right angles. The resulting diffraction

Fig 76
Spectrometer

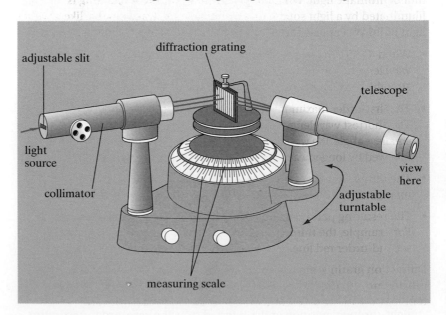

pattern is observed through a small telescope. We know the spacing, d, between the lines on the diffraction grating. The angle at which each order occurs can be carefully measured. The wavelength is then found from the equation $d \sin \theta = n\lambda$.

Example

Light from a monochromatic source shines on to a diffraction grating which has 375 lines per millimetre. The first-order diffraction maxima is found at an angle of 12°.

(a) Calculate the wavelength of the light.

(b) How many diffraction orders are visible?

Answer

(a) Using $n\lambda = d \sin \theta$, the spacing of the diffraction grating is

$$d = 1/375\,000\,\text{m}$$
$$= 2.67 \times 10^{-6}\,\text{m}$$

For $n = 1$, this gives $\lambda = 2.67 \times 10^{-6}\,\text{m} \times \sin 12°$
$$= 555 \times 10^{-9}\,\text{m} = 555\,\text{nm}$$

(b) The order of the maximum is given by $n = d/\lambda \times \sin \theta$. This is a maximum when $\theta = 90°$, i.e. when $\sin \theta = 1$:

$$n = \frac{d}{\lambda} = \frac{2.67 \times 10^{-6}\,\text{m}}{555 \times 10^{-9}\,\text{m}} = 4.81$$

Therefore the highest order to be visible is the fourth. There will be nine orders visible altogether (2 × four orders plus the zero-order).

Examiners' Notes

Remember that d is the distance between adjacent slits. Questions usually give the number of lines per metre, N. You need to calculate $1/N$ to find d.

Spectral analysis

So far we have only considered the diffraction pattern produced by monochromatic light. What happens when a diffraction grating is illuminated by a light source that contains many wavelengths, like a white-light bulb? We can use the equation $d \sin \theta = n\lambda$, to answer this.

• When $\theta = 0°$, the path difference is zero. All waves, no matter what the wavelength, will arrive in phase. If the light source is white, then the straight-through beam will also be white.

• The first-order maximum occurs at an angle that depends on wavelength. The shortest wavelength waves will have an interference maximum at the smallest angle. We would expect to see violet and blue maxima first followed by longer wavelengths up to red at the largest angles.

• The diffraction grating has the effect, similar to a prism, of splitting light up into its various colours. Each order now becomes a **spectrum**.

• The resulting pattern can be complex. Adjacent spectra often overlap. For example, the third-order blue line could lie at a similar angle to the second-order red line.

Diffraction gratings are useful for analysing the light from stars, enabling astronomers to determine what elements are present, what the surface temperature of the star is and even how fast the star is moving.

Examination preparation

How Science Works

Data and formulae

Practice exam-style questions

Answers, explanations, hints and tips

How Science Works

As well as understanding the physics in this unit, you are expected to develop an appreciation of the nature of science, the way that scientific progress is made and the implications that science has for society in general. GCSE and A-level science syllabuses refer to these areas as '*How Science Works*'.

The *How Science Works* element of your course, which also includes important ideas about experimental physics, may be assessed in the written examination papers as well as in the internally assessed unit, the Investigative Skills Assessment or ISA. The concepts included in *How Science Works* may be divided into several areas.

Theories and models

Physicists use theories and models to attempt to explain their observations of the universe around us. These theories or models of the real world can then be tested against experimental results. Scientific progress is made when experimental evidence is found that supports a new theory or model.

You are expected to be aware of historical examples of how scientific theories and models have developed and how this has changed our knowledge and understanding of the physical world. For example, Isaac Newton originally theorised that light was a stream of particles, which he called 'corpuscles'. These corpuscles could bounce off certain materials, thus explaining the reflection of light. Newton also suggested that the corpuscles were attracted by a force as they approached a surface such as water or glass. This changed the velocity of the particles, and Newton used this idea to explain refraction. Alternative models of light, which considered light as a wave motion, were developed by both Christian Huygens and Robert Hooke. These models also explained the phenomena of reflection and refraction. There was, however, one crucial difference. Newton's theory predicted that light would travel faster in a denser medium, such as water or glass, whereas wave theory predicted that the speed of light would be less in these materials. But scientists in the late 17th century had no way of measuring the speed of light so that these contradictory predictions might be tested. In the absence of convincing experimental data, Newton's theory was generally accepted, principally because of his reputation.

Experimental results can of course disprove a theory that was previously accepted and so it was with Newton's corpuscular theory of light. In the early 1800s, Thomas Young demonstrated interference effects in light, which cast doubt on Newton's corpuscular theory. In 1816, the French scientist Augustin Fresnel published a mathematical theory that described light as a wave. The French mathematician Siméon-Denis Poisson pointed out that if Fresnel was correct, then, in some circumstances, there should be a bright spot at the middle of a shadow:

'Let parallel light impinge on an opaque disk, the surrounding being perfectly transparent. The disk casts a shadow, of course, but the very centre of the shadow will be bright.'

Experiments soon confirmed this, and the wave theory of light was accepted – at least until 1905, when Albert Einstein wrote a paper about the photoelectric effect. Einstein's explanation involved describing light as a particle, now called the photon (see Unit 1).

You should know the meaning of the terms 'hypothesis' and 'prediction'.

- A **hypothesis** is a tentative idea or theory, or explanation of an observation.

- A **prediction** from a hypothesis or theory is a forecast that can be tested by experiment.

If a reliable experiment does not support a hypothesis, then the hypothesis is likely to be abandoned or modified. Hypotheses are not usually widely accepted until the experimental results have been repeated by a number of independent scientists. It may take many experimental tests until a set of hypotheses become accepted as a scientific theory (see the figure below). Even then, a scientific theory is always capable of being later refuted if compelling experimental evidence suggests that a new explanation is necessary.

The stages of scientific research

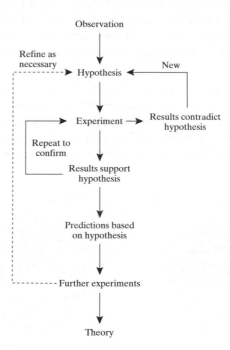

Experimental techniques

You are expected to develop the skills of experimental planning, observation, recording and analysis. These skills will mainly be assessed in the practical coursework, the ISA, or by the externally-marked practical assignment, the EMPA. This section contains some general advice for carrying out experimental work in physics.

When you plan an experiment you need to be able to identify the dependent, independent, and control variables that are involved.

- The **independent variable** is the physical quantity that you deliberately change.

- The **dependent variable** changes as a result of this.

For example, if you are asked to investigate how the length of a piece of metal wire affects its electrical resistance, the length is the independent variable, and the electrical resistance is the dependent variable. Any other variables that may have an impact on the outcome need to be controlled so that the conclusions of the experiment are clear. These are known as the **control variables**. In the example of the wire, two of the control variables are the cross-sectional area of the wire and its temperature.

You will need to select appropriate apparatus, including measuring instruments of a suitable precision and accuracy. These two terms are often confused.

- **Accuracy** refers to how close the reading is to the accepted value.
- **Precision** refers to the number of significant figures that the measurement is made to.

For example, an electronic balance that gives an answer to 0.01 g, e.g. 3.24 g, is capable of more precise measurements than a balance that measures in grams only, e.g. 3 g. If several readings of the same measurement are closely grouped together, the readings are said to be **precise**. If the readings agree with a known mass, they are said to be **accurate**. The analogy of rifle shots at a target may be used to differentiate between accuracy and precision (see the figure below).

Figure A shows accurate shooting, since the bullets (or readings) are close to the centre of the target. But the shooting is not precise, since the bullets are widely scattered. Figure B shows precise shooting, since the bullets (readings) are closely grouped, but not accurate, since the bullets (readings) are not close to the centre of the target (accepted value).

A accurate   **B** precise

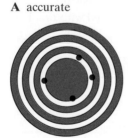 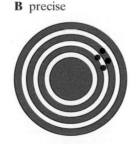

When choosing your apparatus you should be aware that ICT can be used to assist with the collection and analysis of experimental data. This may mean using a suitable sensor, attached to a data logger, to take the readings and then a spreadsheet to help to analyse them. For example, suppose that you wanted to investigate the current surge that passes through the filament of an incandescent light bulb when it is first turned on. The light bulb reaches its operating temperature very quickly, so the readings need to be taken rapidly. A current sensor attached to a data logger could take the readings at the required rate. Conversely, sometimes the readings need to be taken over a long time period, for example if you wanted to investigate the cooling of a house overnight. A temperature sensor and a data logger would make the job somewhat less tedious!

Planning an experiment also means being aware of any risks to health and safety, and taking any necessary precautions. For example, you need to wear protective goggles when stretching a metal wire.

You also need to plan to reduce experimental errors. There are two types of error: systematic and random.

- **Systematic errors** cannot be reduced by repeating the measurement; for example, using an electronic balance which is not zeroed would lead to a systematic error.

- **Random errors** occur when taking readings, such as the timing errors when measuring the period of a pendulum. These can be reduced by repeating the readings and finding the mean, since the errors are random and may fluctuate above and below an average reading.

It is important to identify the percentage uncertainty associated with each reading. For example, if a length is measured using a ruler with millimetre divisions the reading in centimetres may be given to a precision of ± 0.1 cm. A reading of 25.2 ± 0.1 cm has a percentage uncertainty of $(^{0.1}/_{25.2}) \times 100\% = 0.4\%$.

The percentage uncertainty can be reduced by using a more precise measuring device, such as a micrometer or vernier callipers instead of a ruler. The percentage uncertainty can also be reduced by increasing the size of the quantity to be measured. For example, to measure the thickness of a sheet of paper, you could measure the thickness of 100 sheets of paper, and then divide by 100 to find the value for a single sheet. When timing the period of a pendulum it is good practice to time a number of oscillations, say 10, and then divide by 10.

When you tabulate your readings you should ensure that the columns are headed with the quantity *and* the unit that it is measured in. It is good practice to also include the uncertainty associated with that reading in the heading, as in the table here.

Length/m ± 0.001 m	Time/s ± 0.1 s
0.023	2.3
0.031	3.1
0.04*	4.0

* see below

You should always quote figures in your results table to the appropriate degree of precision, and be consistent. The length reading in the last row of the table isn't correct – it should read 0.040.

The most significant uncertainty in the readings taken in an experiment may determine whether you can draw a reliable conclusion or not. You can also test reliability by repeating the experiment a number of times and comparing your results. In practice scientists share their results through publication. Reliability is tested by other scientists who try to replicate the work, or check the results using a different experimental method.

When you plot your results on a graph, you should choose a scale so that the range of your points covers at least half of the graph paper, in both the

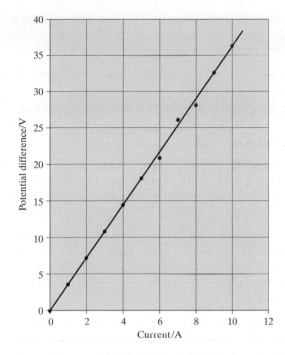

Graph to show how the potential difference across a wire affects the current through it

x- and *y*-directions. Choose scales that have divisions that are easy to interpret accurately, i.e. divisions in multiples of 2, 5 or 10 rather than 3 or 7.

You must label each axis of the graph with the quantity and the unit; conventionally this is done using a solidus, for example length/m. The best-fit line or curve should be drawn so as to minimise the total distance of points from the line. In practice this will mean that some points fall above the line and some below (see the graph here). Any results that fall outside the expected range of values (anomalous values) can be ignored when choosing the best line, but you should identify these anomalous readings on your graph. You should also repeat these readings whenever possible and try to explain why they do not follow the trend. Finally, when measuring the gradient of a line, choose a large section of the graph so as to reduce errors.

Applications and implications of science

Scientific advances have greatly improved the quality of life for the majority of people, and developments in technology, medicine and materials continue to further these improvements at an increasing rate. However, technologies themselves pose significant risks that have to be balanced against the benefits. For example, the way in which electrical energy is generated poses many questions. Nuclear power is capable of generating large amounts of energy and does not emit carbon dioxide, but it does produce radioactive waste which needs to be stored safely for thousands of years.

Scientific findings and technologies enable advances to be made that have potential benefit for humans; however, the scientific evidence available to policy makers may be incomplete. Political decision-makers are influenced by many things, including prior beliefs, vested interests, public opinion and the media, as well as by expert scientific evidence.

Data and formulae

FUNDAMENTAL CONSTANTS AND VALUES

Quantity	Symbol	Value	Units
speed of light in vacuo	c	3.00×10^8	m s^{-1}
permeability of free space	μ_0	$4\pi \times 10^{-7}$	H m^{-1}
permittivity of free space	ε_0	8.85×10^{-12}	F m^{-1}
charge of electron	e	-1.60×10^{-19}	C
the Planck constant	h	6.63×10^{-34}	J s
gravitational constant	G	6.67×10^{-11}	$\text{N m}^2 \text{kg}^{-2}$
the Avogadro constant	N_A	6.02×10^{23}	mol^{-1}
molar gas constant	R	8.31	$\text{J K}^{-1} \text{mol}^{-1}$
the Boltzmann constant	k	1.38×10^{-23}	J K^{-1}
the Stefan constant	σ	5.67×10^{-8}	$\text{W m}^{-2} \text{K}^{-4}$
the Wien constant	α	2.90×10^{-3}	m K
electron rest mass (equivalent to $5.5 \times 10^{-4}\,\text{u}$)	m_e	9.11×10^{-31}	kg
electron charge–mass ratio	e/m_e	1.76×10^{11}	C kg^{-1}
proton rest mass (equivalent to $1.00728\,\text{u}$)	m_p	1.67×10^{-27}	kg
proton charge–mass ratio	e/m_p	9.58×10^7	C kg^{-1}
neturon rest mass (equivalent to $1.00867\,\text{u}$)	m_n	1.67×10^{-27}	kg
gravitational field strength	g	9.81	N kg^{-1}
acceleration due to gravity	g	9.81	m s^{-2}
atomic mass unit (1 u is equivalent to 931.3 MeV)	u	1.661×10^{-27}	kg

ASTRONOMICAL DATA

Body	Mass/kg	Mean radius/m
Sun	2.0×10^{30}	7.0×10^8
Earth	6.0×10^{24}	6.4×10^6

GEOMETRICAL EQUATIONS

arc length $= r\theta$

circumference of circle $= 2\pi r$

area of circle $= \pi r^2$

area of cylinder $= 2\pi rh$

volume of cylinder $= \pi r^2 h$

area of sphere $= 4\pi r^2$

volume of sphere $= \dfrac{4}{3}\pi r^3$

PARTICLE PHYSICS

Rest energy values

Class	Name	Symbol	Rest energy /MeV
photon	photon	γ	≈ 0
lepton	neutrino	ν_e	≈ 0
		ν_μ	≈ 0
	electron	$e^\pm$	0.510999
	muon	$\mu^\pm$	105.659
mesons	pion	$\pi^\pm$	139.576
		π^0	134.972
	kaon	$K^\pm$	493.821
		K^0	497.762
baryons	proton	p	938.257
	neutron	n	939.551

Properties of quarks (antiquarks have opposite signs)

Type	Charge	Baryon number	Strangeness
u	$+\frac{2}{3}$	$+\frac{1}{3}$	0
d	$-\frac{1}{3}$	$+\frac{1}{3}$	0
s	$-\frac{1}{3}$	$+\frac{1}{3}$	-1

Properties of leptons

	lepton number
particles: $e^-, \nu_e; \mu^-, \nu_\mu$	$+1$
antiparticles: $e^+, \bar{\nu}_e; \mu^+, \bar{\nu}_\mu$	-1

Photons and energy levels

photon energy $\qquad E = hf = hc/\lambda$

photoelectricity $\qquad hf = \phi + E_{k(max)}$

energy levels $\qquad hf - E_1 - E_2$

de Broglie wavelength $\qquad \lambda = \dfrac{h}{p} = \dfrac{h}{mv}$

ELECTRICITY

current and p.d. $\qquad I = \dfrac{\Delta Q}{\Delta t} \qquad V = \dfrac{W}{Q} \qquad R = \dfrac{V}{I}$

e.m.f. $\qquad \varepsilon = \dfrac{E}{Q} \qquad \varepsilon = I(R + r)$

resistors in series $\qquad R = R_1 + R_2 + R_3 + \dots$

resistors in parallel $\qquad \dfrac{1}{R} = \dfrac{1}{R_1} + \dfrac{1}{R_2} + \dfrac{1}{R_3} + \dots$

resistivity $\qquad \rho = \dfrac{RA}{L}$

power $\qquad P = VI = I^2R = \dfrac{V^2}{R}$

alternating current $\qquad I_{rms} = \dfrac{I_0}{\sqrt{2}} \qquad V_{rms} = \dfrac{V_0}{\sqrt{2}}$

MECHANICS

moments $\qquad$ moment $= Fd$

velocity and acceleration $\quad v = \dfrac{\Delta s}{\Delta t} \qquad a = \dfrac{\Delta v}{\Delta t}$

equations of motion $\qquad v = u + at \qquad s = \dfrac{(u+v)}{2}t$

$$v^2 = u^2 + 2as \qquad s = ut + \dfrac{at^2}{2}$$

force $\qquad F = ma$

work, energy and power $\quad W = Fs\cos\theta$

$$E_k = \tfrac{1}{2}mv^2 \qquad \Delta E_p = mg\Delta h$$

$$P = \dfrac{\Delta W}{\Delta t} \qquad P = Fv$$

efficiency $= \dfrac{\text{useful output power}}{\text{input power}}$

MATERIALS

density $\qquad \rho = \dfrac{m}{V} \qquad\qquad$ Hooke's Law $\quad F = k\Delta L$

Young modulus $= \dfrac{\text{tensile stress}}{\text{tensile strain}} \qquad$ tensile stress $= \dfrac{F}{A}$

$$\text{tensile strain} = \dfrac{\Delta L}{L}$$

energy stored $\qquad E = \tfrac{1}{2}F\Delta L$

WAVES

wave speed $\qquad c = f\lambda \qquad\qquad$ period $\quad T = \dfrac{1}{f}$

fringe spacing $\qquad w = \dfrac{\lambda D}{s} \qquad$ diffraction grating $\quad d\sin\theta = n\lambda$

refractive index of a substance $s \qquad n = \dfrac{c}{c_s}$

for two different substances of refractive indices n_1 and n_2,

law of refraction $\qquad n_1\sin\theta_1 = n_2\sin\theta_2$

critical angle $\qquad \sin\theta_c = \dfrac{n_2}{n_1}$ for $n_1 > n_2$

Practice exam-style questions

1 (a) (i) The *moment* of a force is its turning effect about a given point. Explain how the *moment* is calculated.

_____

_____ 2 marks

(ii) Explain what is meant by the *principle of moments*.

_____

_____ 1 mark

(iii) Use the principle of moments to determine the maximum load that can be lifted by the crane shown below. You may ignore the weight of the crane. Give your answer in newtons.

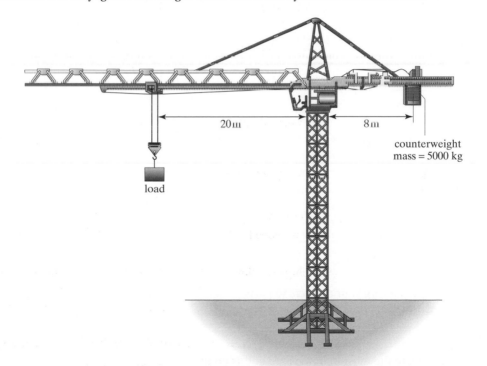

counterweight
mass = 5000 kg

load

20 m

8 m

_____

_____

_____ 3 marks

(b) State one adjustment that could be made to the crane to allow it to lift a heavier load.

_____ 1 mark

Total Marks: 7

2   A bungee jumper jumps from a platform. His feet are tied to a strong rubber cord of length 30 m. The mass of the bungee jumper is 80 kg.

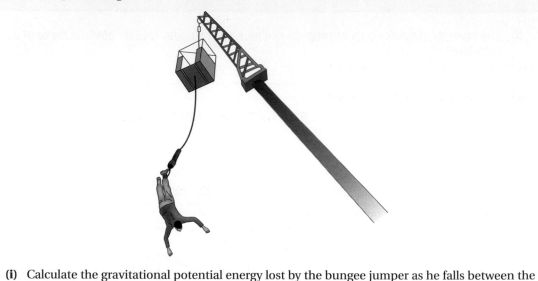

(a) (i) Calculate the gravitational potential energy lost by the bungee jumper as he falls between the platform and the point when the cord reaches its natural (unstretched) length.

_____

_____ 2 marks

  (ii) Calculate the jumper's kinetic energy at that point.

_____ 1 mark

  (iii) What assumption have you made in your answer to (ii)? Explain why this assumption is justified.

_____

_____ 2 marks

(b) The bungee cord stretches until it is twice its natural length and brings the jumper to rest.

  (i) By considering your answers to parts (a)(i) and (ii), calculate how much energy is now stored as elastic strain energy in the cord.

_____ 1 mark

  (ii) Calculate the tension in the cord at this point.

_____

_____ 2 marks

  (iii) Explain why this tension is greater than the jumper's weight.

_____

_____ 1 mark

Total Marks: 9

3 **(a)** Explain the difference between a vector and a scalar quantity.

_____

_____ 2 marks

(b) Give one example of a scalar quantity. _____

Give one example of a vector quantity. _____ 2 marks

(c) A shot putter throws the shot at a speed of $12\,\mathrm{m\,s^{-1}}$ at an angle to the horizontal of $35°$.

35°

(i) Calculate the vertical component of the velocity.

_____

_____ 2 marks

(ii) Calculate the maximum height, above the release point, that the shot will reach.

_____

_____

_____ 3 marks

(iii) How long will it take to reach maximum height?

_____

_____ 2 marks

Total Marks: 11

4

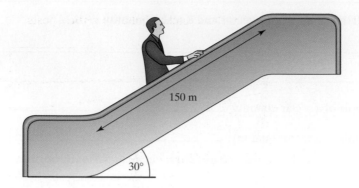

150 m

30°

An escalator travels at $0.5\,\text{m s}^{-1}$ and rises at an angle of 30° to the horizontal. The escalator is 150 m long.

(a) A man of mass 80 kg travels on the escalator. How much work is done by the escalator in raising the man from the bottom to the top of the escalator?

_____

_____ 2 marks

(b) **(i)** The escalator can lift a maximum total mass of 10 000 kg at this speed. What is its useful maximum output power?

_____

_____ 2 marks

(ii) The input power to the escalator is 45 kW. What is the efficiency of the escalator when it is lifting its maximum load?

_____

_____ 1 mark

(c) Suggest why the efficiency of the escalator is less than 100%.

_____

_____ 2 marks

(d) The average efficiency of the escalator, calculated over the course of a day, is much less than the answer to (b)(ii). Discuss why this might be.

_____

_____

_____ 2 marks

Total Marks: 9

5 Some motorway crash barriers use steel cables stretched between vertical posts.

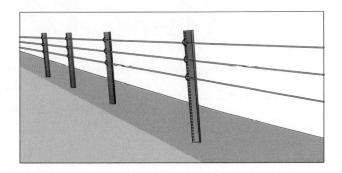

The barrier above is made from three cables. Each cable is made from three wires twisted together. Each wire has a cross-sectional area of 280 mm² and is under an initial tension of 5 kN.

(a) **(i)** Calculate the tensile stress in each wire.

_____

_____ 2 marks

(ii) The value of the Young modulus for steel is 200 GPa. Calculate the extension produced in a 100 m section of wire as it is put under its initial tension of 5 kN.

_____

_____

_____ 3 marks

(b) The yield stress of a wire is 500 MPa.

(i) Explain what is meant by yield stress.

_____

_____ 1 mark

(ii) What is the maximum force that can be applied to a *cable* before it yields?

_____

_____

_____ 2 marks

Total Marks: 8

6

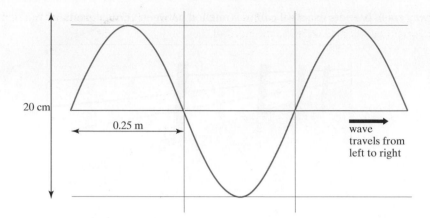

The diagram above shows a *transverse progressive* wave travelling from left to right on a string.

(a) **(i)** Explain what is meant by a *transverse* wave.

_____ 1 mark

(ii) Explain what is meant by a *progressive* wave.

_____ 1 mark

(b) **(i)** What is the wavelength of the wave? _____ m

(ii) What is the amplitude of the wave? _____ m

(iii) The frequency of the wave is 10 Hz. Calculate the speed of the wave.

_____ $m\,s^{-1}$ 3 marks

(c) Sketch the wave as it would look 0.025 s later.   2 marks

(d) A *standing wave* is now produced on the string by fixing one end, so that the reflected wave interferes with the original wave. The diagram below shows a standing wave at one moment in time. The wavelength and frequency of the wave are the same as in part (b).

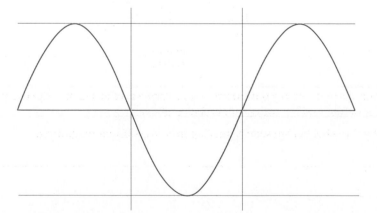

Sketch the wave as it would look 0.05 s later.   2 marks

Total Marks: 9

7 A triangular prism is made of glass of refractive index 1.45. A ray of monochromatic red li~ the side of the prism at an angle of 30° to the normal.

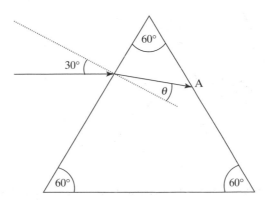

(a) Calculate the angle of refraction, θ.

_____

_____ 3 marks

(b) Calculate the critical angle between the glass and the air.

_____ 2 marks

(c) Draw the path of the light ray after it reaches point A. Mark in the values of the angles.   3 marks

Total Marks: 8

8 A diffraction grating is illuminated with monochromatic laser light of wavelength 633 nm. The diffraction grating has 900 lines per mm.

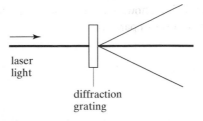

The experiment is carried out in a darkroom. Some smoke is released into the room to allow the beams to be photographed. A sketch of the photograph is shown above.

(a) Explain why three beams are seen emerging from the diffraction grating.

_____

_____

_____

_____ 3 marks

(b) Using a calculation, show that no more than three beams can be seen at this wavelength with this grating.

_____

_____

_____

_____

_____ 3 marks

(c) Describe and explain what you would see if the red laser light was replaced with a beam of white light.

_____

_____

_____

_____

_____ 3 marks

Total Marks: 9

Answers, explanations, hints and tips

Question	Answer		Marks
1 (a)(i)	The moment of a force is calculated using: moment = (magnitude of the) force $\times$ perpendicular distance of the (line of action) of the force from the point	(2)	
(ii)	The principle of moments states that a body can only be in equilibrium if the sum of the moments about any point is zero (i.e. sum of the clockwise moments about any point is equal to the sum of the anticlockwise moments about that point).	(1)	
(iii)	Clockwise moment = $5000 \times 9.8 \times 8 = 392\,000\,\text{N m}$ So for equilibrium, anticlockwise moment = $392\,000\,\text{N m}$ Load $\times 20 = 392\,000\,\text{N m}$, so load = $392\,000/20 = 19\,600\,\text{N}$	(2) (1)	6
1 (b)	*Either* move the counterweight towards the end of the horizontal beam (away from the vertical support) *or* increase the weight (or mass) of the counterweight *or* move the load nearer to the vertical support.	(1)	1
		Total 7	
2 (a)(i)	Use $\Delta E_{p} = mg\Delta h$ $= 80 \times 9.81 \times 30 = 23\,500\,\text{J}$ (lose 1 mark if more than 3 s.f. given; lose 1 mark for missing or incorrect unit)	(1) (1)	
(ii)	$\Delta E_{k} = \Delta E_{p} = 23\,500\,\text{J}$ (must be the same answer as (a) (i), allow error carried forward)	(1)	
(iii)	All of the jumper's (loss in) potential energy has been transferred to kinetic energy. This is justified because air resistance will be negligible (at these low speeds/for a compact/dense object).	(1) (1)	5
2 (b)(i)	Energy is twice the answer to (a) (i) since he has fallen twice as far, so $47\,000\,\text{J}$	(1)	
(ii)	Use $E = \frac{1}{2}F\Delta l$, so $F = 2E/\Delta l$ $= 2 \times 47\,000/30 = 3130\,\text{N}$ (to 3 s.f.)	(1) (1)	
(iii)	The tension is higher than the jumper's weight because the jumper is not in equilibrium (it will accelerate the jumper upwards).	(1)	4
		Total 9	
3 (a)	A scalar quantity is defined by its magnitude, whereas a vector quantity is defined by its magnitude *and* its direction.	(1) (1)	2
3 (b)	Scalar: one of, for example, energy, speed, temperature, mass. Vector: one of, for example, velocity, force, displacement, acceleration.	(1) (1)	2
3 (c)(i)	Vertical component = $v\sin\theta = 12 \times \sin 35°$ $= 6.88\,\text{m s}^{-1}$	(1) (1)	

Question	Answer		Marks
3 (c)(ii)	When the shot reaches maximum height, the vertical velocity will be zero.	(1)	
	Use $v^2 = u^2 + 2as$	(1)	
	to give $s = -u^2/2a = -6.88^2/-9.81 \times 2 = 2.41\,\text{m}$	(1)	
	(or by use of energy)		
(iii)	Use $v = u + at$, so $t = -u/a = -6.88/-9.81$	(1)	
	$= 0.70\,\text{s}$	(1)	7
			Total 11
4 (a)	Work done $= Fs\cos\theta = 80 \times 9.81 \times 150 \times \cos 60°$	(1)	
	$= 58\,800\,\text{J}$	(1)	
	($60°$ is the angle between the weight of the man and the distance travelled)		
	Or work done $= E_p$ gained by man $= mg\,\Delta h$		
	$= 80 \times 9.81 \times 150 \times \sin 30°$	(1)	
	$= 58\,800\,\text{J}$	(1)	2
4 (b)(i)	Power $= Fv\cos\theta = 10\,000 \times 9.81 \times 0.5\cos 60°$	(1)	
	$= 24.5\,\text{kW}$	(1)	
	or		
	Time to lift mass $= d/v = 150/0.5 = 300\,\text{s}$	(1)	
	Power $=$ work done / time $= (10\,000 \times 9.81 \times 150 \times \cos 60°)/300$		
	$= 24.5\,\text{kW}$	(1)	
(ii)	Efficiency $=$ useful power out /total power in $= 24.5/45 = 54.4\%$	(1)	3
4 (c)	Energy is lost as heat due to friction/work done against friction	(1)	
	and as heat in electrical wires.	(1)	2
	Or energy is needed to lift escalator itself which has significant weight (though this is balanced by an equal weight falling because the escalator is a loop).		
4 (d)	The escalator sometimes runs at less than full capacity		
	(i.e. sometimes there are fewer people on the escalator).	(1)	
	This reduces the useful output power. /The energy losses still occur.	(1)	2
			Total 9
5 (a)(i)	Stress $=$ force/area $= 5000\,\text{N}/280 \times 10^{-6}\,\text{m}^2$	(1)	
	$= 17.9\,\text{MPa}$	(1)	
(ii)	Young modulus $E =$ stress/strain, so strain $=$ stress/E	(1)	
	$= 17.9 \times 10^6/200 \times 10^9 = 8.93 \times 10^{-5}$	(1)	
	Extension $=$ strain $\times$ original length $= 8.93 \times 10^{-5} \times 100\,\text{m}$		
	$= 8.93\,\text{mm}$	(1)	5
5 (b)(i)	The yield stress is the stress at which the extension of the wire changes from elastic deformation to plastic deformation, causing it to permanently extend.	(1)	

Question	Answer		Marks
5 (b)(ii)	Maximum force on one wire = yield stress of wire × cross-sectional area of wire		
	$= 500 \times 10^6 \times 280 \times 10^{-6} = 140 \, \text{kN}$	(1)	
	But there are three wires in a cable, so maximum force possible		
	$= 140 \times 3 = 420 \, \text{kN}$	(1)	3
			Total 8
6 (a)(i)	A transverse wave has oscillations which are at right angles to the direction of travel of the wave.	(1)	
(ii)	A progressive wave transfers energy in the direction of propagation (travel).	(1)	2
6 (b)(i)	0.5 m	(1)	
(ii)	0.1 m	(1)	
(iii)	$v = f\lambda = 10 \times 0.5 = 5 \, \text{m s}^{-1}$	(1)	3
6 (c)	The frequency of the wave is 10 Hz so its period is 0.1 s. 0.025 s is $\frac{1}{4}$ of the wave period. after 0.025 s	(2)	2
6 (d)	after 0.05 s	(2)	2
			Total 9

Question	Answer		Marks
7 (a)	$\sin 30°/\sin\theta = 1.45$	(1)	
	$\sin\theta = \sin 30°/1.45 = 0.34$	(1)	
	$\theta = 20.2°$	(1)	3
7 (b)	$\sin c = 1/n = 1/1.45 = 0.70$	(1)	
	$c = 43.6°$	(1)	2
7 (c)	Angle of incidence = 39.8°	(1)	
	Less than critical angle, so ray is refracted out of prism, away from normal.	(1)	
	Angle of refraction = 68.1°	(1)	3
			Total 8
8 (a)	The central (straight-through) beam is where all the waves from the grating are in phase, because there is a zero path difference.	(1)	
	The beams above and below the central beam are caused by light waves from adjacent slits in the grating having a path difference of one wavelength.	(1)	
	The waves from adjacent slits are therefore in phase in those directions.	(1)	3
8 (b)	Using $d\sin\theta = n\lambda$, the maximum value for θ is 90°, for which $\sin\theta = 1$.	(1)	
	Then $n = d/\lambda$		
	$d = 1/900\,000 = 1.11 \times 10^{-6}\,\text{m}$ and $\lambda = 6.33 \times 10^{-7}\,\text{m}$, giving $n = 1.76$	(1)	
	n has to be an integer, so the largest possible value is 1. There is one first-order beam either side of the central beam, so three altogether.	(1)	3
8 (c)	The straight-through beam would be white,	(1)	
	as all wavelengths would be in phase.	(1)	
	Either side of the straight-through beam there would be a spectrum,	(1)	
	with violet/blue light closer to the middle and red at a larger angle.	(1)	
	Using $d\sin\theta = n\lambda$, with $n = 1$ (first-order maximum), the smallest wavelengths will be in phase at smaller values of $\sin\theta$, (or θ).	(1)	
		(any 3)	3
			Total 9

Glossary

absolute refractive index (n)	property of an optical material equal to the ratio: speed of light in a vacuum / speed of light in the material
acceleration	the rate of change of velocity: change in velocity / time taken; unit $m\,s^{-2}$
acceleration due to gravity	the rate at which all objects accelerate under gravity if air resistance is neglected; also known as the acceleration of free-fall; on Earth it is usually taken as $9.81\,m\,s^{-2}$, but it varies slightly from place to place
amplitude	the maximum height of a wave, or the largest displacement from equilibrium
antinode	a point on a stationary (standing) wave where the amplitude is a maximum
in antiphase	two points on a wave, or points on two waves, are in antiphase if their vibrations are $180°$ out of phase with each other
breaking stress (or ultimate tensile stress)	the maximum stress (force per unit area) that a material can withstand before it breaks
brittle	a brittle material fractures before it has undergone plastic deformation
centre of gravity	the point at which the weight of an object can be taken to act; an object will balance if it is supported at its centre of gravity
centre of mass	the point at which the mass of an object can be taken to be concentrated; in a uniform gravitational field, this is the same point as the centre of gravity
cladding	a layer of glass (or plastic) that surrounds the central core of an optical fibre
coherent	two or more waves that have a fixed phase difference are said to be coherent
component	a vector can be split up into perpendicular components; the vertical component is that part of the vector that acts in a vertical direction
compression	an object in compression is under the influence of forces that tend to squash it
coplanar forces	a two-dimensional system of forces that all act in the same plane; they can be drawn on a piece of paper
couple	two equal forces that act in opposite directions on an object so as to cause rotation
critical angle	the minimum angle of incidence at an optical boundary at which total internal reflection occurs
density	the amount of mass per unit volume; symbol ρ; unit $kg\,m^{-3}$
diffraction	the spreading of waves through an aperture or round an obstacle
diffraction grating	a series of closely spaced parallel slits through which light can diffract; used to create spectra
displacement	a vector describing the difference in position of two points
drag	a resistive force, such as air resistance, which acts to oppose motion in a fluid
ductility	the ability of materials to show extended plastic deformation and become elongated under tension; a ductile metal can be drawn out into wires

efficiency	the ratio: useful energy transferred (or work done) / total energy input; this is always less than 1
elastic behaviour (elasticity)	when a material returns to its original dimensions after a deforming force is removed; it is said to behave elastically
elastic limit	the maximum tensile force at which a material behaves elastically, i.e. returns to its original dimensions after a deforming force is removed
elastic strain energy	the potential energy stored in an elastic material that has been extended
endoscope	a medical device that uses optical fibres to see inside the body
energy	the ability to do work, i.e. move a force through a distance; a scalar quantity, measured in joules (J)
equilibrium	an object is said to be in equilibrium if it is not accelerating
first-order maximum	a point at which the waves passing through a diffraction grating interfere constructively; waves from adjacent slits have a path difference of one wavelength, and so all the waves arrive in phase
free body diagram	a simplified picture of a physical situation which shows all of the relevant forces acting on a body
frequency	the number of waves passing a point in one second, measured in hertz, Hz
friction	a force that acts between surfaces, acting so as to oppose their relative motion
fundamental frequency	the lowest resonant frequency of a vibrating system or standing wave
gravitational potential energy	the energy stored by a mass due to its position in a gravitational field; in a uniform field, the gravitational potential energy of a mass m that is raised by a distance Δh is given by $E_\text{p} = mg\Delta h$
Hooke's Law	law stating that, for an object under tension, such as a wire or a spring, the extension is proportional to the applied force
inertia	an object's resistance to acceleration; for linear motion, this is the mass
instantaneous velocity	the rate of change of displacement, as measured over a very small time interval
interference pattern	a series of maxima (points of constructive interference) and minima (points of destructive interference) in a region where two or more waves overlap
kinetic energy	the energy of a mass m moving at a velocity v; $E_\text{k} = \frac{1}{2}mv^2$
laser	a device that produces a highly monochromatic, coherent, non-diverging light beam
longitudinal wave	a wave that has oscillations parallel to the direction of travel of the wave
moment	the turning effect of a force; the moment of a force about a point is equal to: force × perpendicular distance from the line of action of the force to the point, either clockwise or anticlockwise; unit N m
momentum (p)	property of a moving object equal to its mass, m, multiplied by its velocity, v; $p = mv$; a vector quantity; unit kg m s^{-1}
monochromatic	monochromatic light has a single wavelength
necking	the reduction in diameter of a wire under tension when it is close to its breaking stress

newton	the S.I. unit of force; 1 newton (1 N) is the force that will accelerate a mass of $1\,\text{kg}$ at $1\,\text{m}\,\text{s}^{-2}$
node	a point on a stationary (standing) wave at which the amplitude is zero
optical fibre	a thin strand of glass or plastic which carries light signals
overtone	a vibration with a frequency that is a multiple of the fundamental frequency
parallelogram law	a method for finding the resultant of two vectors
path difference	the difference in the distance travelled by two waves; commonly expressed as a number of wavelengths, λ ; e.g. 'a path difference of $\frac{1}{2}\lambda$'
phase difference	the difference in phase (the position in the cycle) of two waves, expressed in degrees or radians; a phase difference of $180°$, or π radians, is equivalent to a path difference of $\frac{1}{2}\lambda$
in phase	two waves are in phase if they are at the same point in their cycle at the same time
plastic behaviour	when a material is permanently deformed, even after the applied force is removed
polarised	a transverse wave that is constrained to vibrate in one direction only is said to be polarised
power	the rate at which energy is transferred or the rate at which work is done, measured in joules per second, or watts, W
principle of conservation of energy	law stating that the total energy of a closed system is constant
principle of moments	law stating that if an object is in equilibrium, the sum of the clockwise moments about any point must equal the sum of the anticlockwise moments about that point
principle of superposition	law stating that when two similar waves overlap, the total disturbance caused is the vector sum of the individual disturbances
progressive wave	a wave that transfers energy in the direction of the wave travel
rarefaction	a region of lower pressure or density in a longitudinal wave
refraction	the change in direction of a wave as it crosses a boundary between two mediums in which its speed differs
refractive index (absolute refractive index) (n)	property of an optical material equal to the ratio: speed of light in a vacuum / speed of light in the material
relative refractive index ($_1n_2$)	the relative refractive index of material 2 relative to material 1 ($_1n_2$) is the ratio: speed of light in medium 1 / speed of light in medium 2
resolution (or resolving)	the splitting up of a vector into components, usually perpendicular
resultant	the sum of two or more vectors, such as forces
scalar	a physical quantity that is fully specified by its magnitude (size); it has no direction associated with it
Snell's Law	law of refraction connecting the angle of incidence and angle of refraction with the (absolute) refractive indices of the materials either side of the boundary: $n_1 \sin\theta_1 = n_2 \sin\theta_2$

spectrometer	a device that uses a diffraction grating to produce spectra or to measure the wavelength of monochromatic light
spectrum	the distribution of wavelengths in a light source
spring constant (k)	the force needed to stretch a spring by unit extension: k = force/extension; unit $N\,m^{-1}$; it is a measure of stiffness of the spring
stationary (or standing) wave	a wave that does not transfer energy in the direction of wave travel; it has stationary points called nodes
stiffness	the resistance to extension of a material under tension
strain (or tensile strain) (ε)	the fractional increase in length of a wire, l, under tension: $\varepsilon = \Delta l / l$; it has no unit
strength	a measure of the force (stress) needed to cause fracture of a material
stress (or tensile stress) (σ)	the force per unit cross-sectional area; $\sigma = F/A$
tensile force	a force acting to cause extension
tension	an object in tension is under the influence of forces which tend to extend it
terminal velocity	the steady velocity reached by a falling object when the drag is equal to the weight
torque	the rotational equivalent of force; torque produces rotational acceleration; unit $N\,m$
total internal reflection	the complete reflection of a light ray at the boundary of two media, when the ray is in the medium with a lower speed of light
transverse wave	a wave that has oscillations perpendicular to the direction of travel of the wave
ultimate tensile stress (or breaking stress)	the maximum stress (force per unit area) that a material can withstand before it breaks
upthrust	the upward force on an object that is submerged in a fluid; it is equal to the weight of fluid displaced
vector	a physical quantity that is specified by its magnitude (size) and its direction
velocity	the rate of change of displacement; velocity = change in displacement / time; unit $m\,s^{-1}$
watt	unit of power, equal to a rate of energy transfer of 1 joule per second
wavelength	the distance between consecutive points on a wave that have identical motion
work	work done = force $\times$ distance moved in the direction of the force
yield point	the minimum stress at which plastic deformation occurs
Young modulus	the stiffness constant of a material, defined by the ratio: tensile stress/tensile strain
zero-order maximum	the central point at which the waves passing through a diffraction grating interfere constructively; waves from adjacent slits have zero path difference and so all the waves arrive in phase

Index